Understanding

FFT

*SECOND EDITION
EXTENSIVELY REVISED*

Applications

A Tutorial for
Students & Working Engineers

Anders E. Zonst

Citrus Press
Titusville, Florida

Publishers Cataloging-in-Publication Data

Zonst, Anders E.
    Understanding FFT Applications—A Tutorial for Laymen, Students, Technicians and Working Engineers/ A.E. Zonst
      p. cm.
      Includes bibliographical references and index
      1. Fourier analysis. 2. Fourier transformations
      3. Digital signal processing   I. Title

QA403.Z658   2004   515.7'23   97-66217
ISBN 0-9645681-4-4

Library of Congress Catalog Card Number: 97-66217

All rights reserved. No part of this book may be reproduced, transmitted or stored by any means, electronic or mechanical, including photocopying, photography, electronic or digital scanning and storing, in any form, without written consent from the publisher. Direct all inquiries to Citrus Press, Box 10062, Titusville, FL 32783.

Copyright ©1997, 2004 by Citrus Press

International Standard Book Number 0-9645681-4-4

Printed in the United States of America 9 8 7 6 5 4 3 2 1

To Maureen

ACKNOWLEDGEMENTS

My old friend John Granville refused to critique the first edition of this book, and when I asked why, he volunteered an impromptu informal criticism (i.e., tirade—sorry I asked). Nonetheless, when I was asked to revise the manuscript for a second edition, John's criticism came back to haunt me. As I read the book in preparation for this revision, all of the minor shortcomings became major, and in the end the task of revision also became major.

Now, except for a few changes which address John's specific criticisms, he has had no direct input to this revision; nonetheless, I must acknowledge he was the primary driving force behind this extensively revised edition. Thanks a lot John.

I would also like to say thanks to the readers who bothered to send comments, but especially to my friend in Seattle, Mr. Larry DeShaw.

I must, once again, give a special thanks to the staff of Citrus Press...I hope you're *adequately* compensated.

Finally, I must acknowledge the support of my wife, who tolerates my moody behavior, and my neglect, when I am involved in these projects. It's one thing to promise support and quite another to live it through the years....

AZ - 30 April 2003

CONTENTS

Prologue vii
Foreword ix

PART I - MATHEMATICAL REVIEW

Chapter 1	About the Sinusoid	1
Chapter 2	The Nature of the Sinusoid	7
Chapter 3	The Mathematically Unique Sinusoid	18

PART II - TRANSIENT CIRCUIT ANALYSIS

Chapter 4	Network Analysis	31
Chapter 5	Circuit Analysis Program	37
Chapter 6	Transient Analysis	47

PART III - THE SPECTRUM ANALYZER

Chapter 7	Speeding Up the FFT	67
Chapter 8	The Inverse PFFFT	83
Chapter 9	The Spectrum Analyzer	91

PART IV - AUDIO APPLICATIONS

Chapter 10	The Sampling Rate	105
Chapter 11	Digital High Fidelity Audio	
	Part 1 - Frequency Related Problems	121
	Part 2 - Amplitude Related Problems	136
Chapter 12	Changing Playback Rate	145

PART V - IMAGE APPLICATIONS

Chapter 13	The Two Dimensional FFT	158
Chapter 14	2-D Transform Familiarization	166
Chapter 15	Fourier Optics	174
Chapter 16	Image Enhancement	185

APPENDICES

Appendix 1.1	Average Value	199
Appendix 4.1	Differentiation & Integration via Xforms	201
Appendix 6.1	Transient Analysis Program	205
Appendix 7.1	Positive Frequency Stretching	206
Appendix 7.2	The Small Array FFT	209
Appendix 9.2	About the Fractional Frequency FFT	211
Appendix 9.3	One Deci-Bel	213
Appendix 9.4	Spectrum Analyzer II	215
Appendix 10.0	Negative Frequencies	221
Appendix 11.1	Oversample Demonstration	225
Appendix 11.2	Oversampling II	227
Appendix 11.3	Digital Audio Analyzer	229
Appendix 12.1	Playback Speedup	235
Appendix 12.2	Music Synthesizer	245
Appendix 13.1	The 2-D FFT Program	253
Appendix 14.1	2-D Functions	258
Appendix 15.1	Fourier Optics Routines	263
Appendix 16.1	Image Enhancement Program	265
Appendix 16.2	Improving Signal/Noise	275
Bibliography		277
Index		278

PROLOGUE

Quietly, a revolution is taking place in one little corner of advanced mathematics—the FFT is coming down from the Ivory Tower and into the practical world. Texts such as Bracewell's, Gaskill's, and Brigham's are obviously steps in this revolution; yet most colleges and universities teach this pretty much as they did 50 and 100 years ago. Seeing this evolution clearly, I try to cut *directly* to the chase, and bring this technology to anyone who has a modest mathematical background. These books will be useful to undergraduate engineers (and, indeed, many high school students); but also to working engineers (and engineering managers) who, in their formal education, were given a glimpse of this technology that did more to intimidate than illuminate.

When one works in technology for a number of years it becomes apparent that, if something works, there's usually a fairly simple reason why. Surely there are *subtle* relationships, but once you discover the *secret*, they're not all *that* difficult...the *secret* lying, more often than not, in grasping a previously unfamiliar concept. These, in fact, are what I try to convey here—fundamental concepts underlying the DFT/FFT.

These are not textbooks. There is much in the textbooks that's deliberately left out. If, however, you understand the concepts presented here you will get much more from a formal course than just memorization of equations you don't understand.

Much more important than the above considerations, however, is the fact that this revolution concerns a way of "seeing" the world we live in. We begin to recognize the two domains of all the variations on Fourier are just different ways of seeing the objectively real world (our perception of sound, for example, is very much a frequency domain phenomenon). By the time we reach the presentation of 2-D transforms this dual aspect of perception and understanding grows clearer, and our grasp of this world we live in grows stronger. This is something that should be understood by more than just a few people with advanced degrees—we should all understand this—it paves the way to comprehension of much that occurs in our daily lives...and hints at much in more obscure areas.

I cannot promise this will be easy, but if you are reading this prologue it's virtually certain you *can* understand this...if you're willing to put in a little work. Precious little is required in the way of prerequisites, except that you be intelligent enough to be interested in this sort of thing.

A. E. Zonst

INTRODUCTION

This companion volume to *Understanding the FFT* assumes the reader is familiar with the material of that book. It's written in five parts covering a range of topics from transient circuit analysis to two-dimensional transforms. It's an introduction to some of the many applications of the FFT, and it's intended for anyone who wants to explore this technology—but primarily for those who have read my first book and now want to experiment with this tool.

The presentation is unique in that it avoids the calculus almost (but not quite) completely. It's a practical "how-to" book, but it also presents a down-to-earth understanding of the technology underlying the discrete Fourier transform (DFT) and the fast Fourier transform (FFT).

This book develops computer programs in BASIC which actually work, and the reader is strongly encouraged to type these into a computer and run them. Some of these computer programs are short routines used to illustrate a relationship or theorem, but mostly they are computer programs that actually perform the application being developed. This is a major feature of this book—unless we understand how to develop, write and apply the computer programs, we never quite understand this subject. The aim is to show how this technology applies to practical problems.

The reader should understand that presentations are frequently started at an elementary level. This is just a technique to establish the foundation for the subsequent discussion, and the material comes quickly to the problem at hand (usually). The book is written in an informal, tutorial style, and should be manageable by anyone with a solid background in high-school algebra, trigonometry, and complex arithmetic. I have tried to include the mathematics that might not be included in a high-school curriculum; so, if you managed to work your way through the first book, you should be able to handle this one.

This book was written because, having presented the FFT in my first book (*Understanding the FFT*) at the level of high school mathematics, it became apparent I might have led many readers into the swamp and abandoned them there. Furthermore, the discovery that this material *could* be presented at this level led me to see the desirability of doing so. Unfortunately, it wasn't clear at the outset what should be included in this book and, consequently, the manuscript sort of grew naturally.... As was pointed out to me later, the lack of organization made the first edition difficult, and I have attempted to correct that sort of mistake here. Most prominent, in this regard, is the improved coherence and readability of the book.

CHAPTER 1

ABOUT THE SINUSOID

1.0 INTRODUCTION

A detailed study of the discrete Fourier transform (DFT) is, as we shall see, primarily a study of sinusoids. Applications of the DFT and FFT concern the systematic manipulation of groups, or *bands* of sinusoids, but we must also study single, individual sinusoids if we hope to get to the bottom of the phenomena that are so intriguing.

The sine function is commonly introduced by Eqn. (1.1). We all know what this means—it simply says we are to take the sine of the *argument* in parentheses. But this equation *hides* the very thing we must

$$f(t) = A_o Sin(2\pi f t + \phi) \quad\quad\quad\quad\quad (1.1)$$

where: A_o = Amplitude
ϕ = Phase
f = frequency

understand here—it hides the *transition* from argument to sine function. To *understand* the intriguing phenomena of Fourier's transform we *must* understand this transition, so let's go back to the beginning and trace the path that gets us to an understanding of this profound relationship.

1.1 SINUSOIDS

The sine of an angle ϕ (the Greek letter phi, pronounced as in "Phi, phi, fo, fum...") is defined using a right triangle, and is the **ratio** of the *side opposite the angle* ϕ to the *hypotenuse* (Fig. 1.1). The cosine of ϕ is the ratio of the adjacent side (i.e., X) to the hypotenuse.

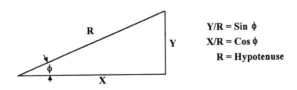

Figure 1.1 - Sine and Cosine Definition

If the hypotenuse is of unit length, the side opposite the angle will be exactly equal to the sine—and the adjacent side will equal the cosine. In Figure 1.2 the unit hypotenuse is also the radius of a circle; so, if we rotate this radius (like the spoke of a wheel), φ will move through all

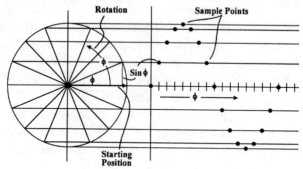

Figure 1.2 - Sinusoid Generation

possible angles. We may plot the successive values of the sine function (i.e., the vertical component of the radius) against the value of φ as the radius rotates. *[Figure 1.2 also illustrates a characteristic of digital (i.e., discrete data) systems. While the angle φ changes continuously, and the sine function is also continuous, we can only sample a finite number of points in a digital system. Note the symmetry of the sampled data—we have 16 (i.e., 2^4) data points. This will be useful in illustrating various points, but 2^N data points (where N is an integer) is completely composed of factors of 2, which provides maximum efficiency in the FFT. It's not imperative, however, that 2^N data points be used in a DFT.]*

The term *sinusoid* includes both the sine and cosine function. The cosine may be generated by starting with the radius in the vertical position (Figure 1.3). It's apparent the *only* difference between Fig. 1.2 and 1.3 is

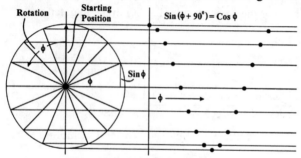

Figure 1.3 - Cosine Generation

Sinusoids

the *phase* (i.e., the starting angle) of the functions. The phase angle can be any value from zero to 2π, of course, and we include all possible phase angles under the term *sinusoid*.

We may also note that, if the rate of rotation is constant, the angle ϕ may be expressed as a function of time. We need only convert this *time* variable into radians, scaled to 2π for each revolution:

$$\phi(t) = 2\pi t/T \qquad (1.2)$$

where: $\phi(t)$ = angle as a function of time (in radians)
            t = time (usually measured in seconds)
            T = period of one revolution

In electrical work the reciprocal of T (i.e., the *frequency* of completing rotations per unit time) is most often used and designated as $f = 1/T$. The expression $2\pi f$ is used so often that we usually shorten Eqn. (1.2) to:

$$\phi(t) = \omega t \qquad (1.3)$$

where: $\omega = 2\pi f$ or $2\pi/T$

We will talk more about this *argument* of sinusoids later, but for now let's continue our present discussion.

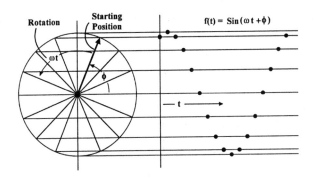

Figure 1.4 - Time-Based Sinusoid

Since the term *sinusoid* includes all possible phase angles, we must include a term for phase in our time-based argument (Fig. 1.4):

$$f(t) = \text{Sin}(\omega t + \phi) \qquad (1.4)$$

where: ϕ = starting (or phase) angle.

Chapter 1

1.2 THE ARGAND DIAGRAM

Figure 1.4 depicts two distinctly different diagrams. The one on the left is for plotting *rotation* and the one on the right depicts a vertical displacement for some *linearly* increasing variable (i.e., *time*). [Note: We have *made* time the common *parameter* between these diagrams, but *time* is not the cause of anything. It's only a common parameter. Equations that are functions of a common parameter are called parametric equations.]

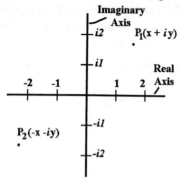

Figure 1.5 - Argand Diagram

The diagram on the left is known as an *Argand diagram*,[1] and we must talk a little about it. The horizontal axis is the *real* axis and the vertical is the *imaginary* axis. The reason for this nomenclature is that, if we allow multiplication by -1 *rotates* a phasor from the positive real axis to the negative (180° in Argand's diagram), then we may hypothesize there is some similar constant (designated as i) which will rotate a phasor *half way* to the negative real axis (i.e., 90° to the imaginary axis). Clearly, then, multiplying by i twice must rotate the phasor to the negative real axis:

$$i \times i = i^2 = -1$$

so that:

$$i = \sqrt{-1} \qquad \text{-----------------------------} \qquad (1.5)$$

but since no *real* number squared produces a negative number we allow this must be an *imaginary* number. Mathematically, the use of $i = \sqrt{-1}$ to *rotate* numbers onto the imaginary axis is completely consistent. A key point here is that Argand diagrams plot *rotational phenomena*—they plot magnitudes and *rotation*. The use of $i = \sqrt{-1}$ allows a consistent mathematical description of rotation—but let's talk about rotation....

---

[1] Jean Robert Argand (1768-1822), who developed a geometrical interpretation of complex numbers...but after Casper Wessel (1745-1818). Long before Argand and Wessel, Rafael Bombelli (born ca. 1530) saw the necessity of recognizing the existence of numbers such as $\sqrt{-1}$ but allowed they were "imaginary numbers." Hamilton (1805-1865) did much to clear up the confusion with complex numbers. Arnold Dresden has suggested removing the stigma of the term *imaginary* by replacing it with *normal*.

Sinusoids

1.3 ABOUT ROTATION

Sinusoids are functions of an angle (i.e., of rotation). That is, the *argument* of a sinusoid ($\omega t + \phi$) describes a quantity of rotation. Rotation is fundamentally different from straight line or *linear* motion. With linear motion we change position, but with rotation *per se* we only change direction. If we stop in the middle of Kansas and ask the way to San Jose, the farmer will consult the weather vane at the top of his barn, rotate 90° counter-clockwise, and point. Note that no distance is implied—distance is moved only if we specify a *radius*, and all circular distance depends on the length of that radius.

The tip of a rotating radius traces-out a circumference of a circle—the length of which is 3.14159265358979323846264338327950028... (i.e., π) times the length of the diameter (i.e., 2 × radius). But, to make this comparison, we must straighten the *circular* circumference into a linear distance—otherwise we're comparing apples to oranges.

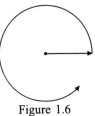

Figure 1.6

We measure rotation itself by specifying an *angle*. We more or less arbitrarily define a full circle of rotation as having 360° and measure all other rotations as fractions of this 360°. There's nothing very special about 360°—we could have chosen 400 "grads" and then each quadrant would contain 100 grads, but these arbitrary definitions don't include much, mathematically, of the cyclical nature of rotation. Once we turn a full circle we've done all there is to do (except to repeat the operation). This "quantum" of rotation has no equivalent in linear motion, and it would be nice if our mathematics included this phenomenon.

Nonetheless, our understanding of rotation is aided by considering the geometry of *circles*. If we divide the circular length of the circumference of any circle by the linear length of the radius of the same circle, the centimeters in the circumference (numerator) will be cancelled by the centimeters in the radius (denominator), and we obtain a dimensionless number 2π...but this isn't quite correct! We have just divided the distance *around* the circle by a *linear* distance, and there must still be a quality of curvature or rotation implicit in this quotient. What we have really done is divide 6.28... *circular centimeters* by one *centimeter*, yielding 2π *circular* or 2π *rotation*. To be consistent, any time we multiply this 2π

rotation constant by a linear distance, we must reintroduce the rotational quality. The term 2π *radians* indicates a length on the circumference *normalized* to the radius—which is a quantity of rotation—an angle.

The term *radian*, then, implies the attribute of rotation, and 2π radians now serves the same function as our arbitrary 360°, and becomes a sort of quantum constant relating rotation to pure numbers. That is, it's apparent that we can take any fraction of these 2π radians just as we took a fraction of 360°. Half a circle contains π radians, a quarter contains $\pi/2$, etc., etc.; however, whole number multipliers serve only to count the number of full rotations. With the 2π *radians* term factored, we may use pure, linear, real numbers to measure the *quantity* of rotation.

So, a 1 cm radius yields a circumference of 2π cm, which is the same number of cm as radians, so the angle in radians is precisely equal to the distance included in the arc along the circumference. Clearly then, if we divide the distance along any circumference by the radius we get the included angle. It's apparent, then, for very small angles, the *sine* of the angle (as well as the tangent) will approximately equal the angle itself. This comes about from the fact that, as we consider smaller and smaller segments of a circle, they look more and more like straight lines (see Figure 1.7). If an astronomer wants to resolve two features on the moon that are 1/4 mile apart, he will know immediately his telescope must be capable of resolving $0.25/250{,}000 = 1 \times 10^{-6}$ radians (actually, astronomers measure small angles in arc minutes and arc seconds, which is obviously better. ;-) Of interest a little later, it's apparent that as the angle approaches zero, $\text{Sin}(\phi)$ approaches zero, but the ratio of $\text{Sin}(\phi)/\phi$ will approach 1.0.

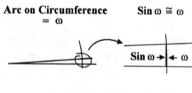

Arc on Circumference = ω $\text{Sin}\,\omega \cong \omega$

$\text{Sin}\,\omega \rightarrow\ \leftarrow \omega$

Figure 1.7

If we multiply 2π radians by a pure dimensionless number, there will be no length associated with the product...it will yield a quantity of pure rotation (i.e., an angle). When we divide the independent variable *t* (i.e., elapsed time since the start of our sinusoid) by the time required to complete one cycle *T* (i.e., the period of a cycle) the dimensions of time cancel, and we obtain a purely numerical proportion, and if we multiply this by 2π radians, we will obtain the angular *position*.

CHAPTER 2

THE NATURE OF THE SINUSOID

A straight line, for example, has simple characteristics—it has a *slope* and an *offset*. As we change the independent variable x the dependent variable y changes proportional to the *slope* in a linear manner. The slope, then, is the dominant characteristic of a straight line, and when we provide for an added constant (i.e., offset) we have said all there is to say about straight lines—we have described their nature. Parabolas, on the other hand, are similar to straight lines except that the slope *changes* at every point on the curve. The differential calculus studies the phenomenon of slope, and as we understand this phenomenon we begin to understand the nature of curves such as parabolas. So, in this sense, what is the *nature* of the curves we call sinusoids.

The most famous function in the mathematics of the calculus is, perhaps, the function whose derivative is equal to itself—the curve whose slope is equal to itself—which we know to as e^x. This unique function describes relationships such as population growth [the rate of growth of a population must be proportional to the number of couples procreating (i.e., the population itself)]. Now, the phenomenon of things changing at a rate proportional to their own magnitude is ubiquitous in nature (e.g., compound interest, charging capacitors, etc., etc.) and so we are not surprised to find that this *exponential function, e^x*, is ubiquitous in our mathematical descriptions of nature. As we grasp the underlying concept, we grasp the nature of *this* function.

The exponential function is of considerable interest to us, for, as we shall see shortly, sinusoids are built on this e^x function. We might anticipate, then, that sinusoids will have characteristics not completely dissimilar to exponential functions (i.e., *their* derivatives might resemble themselves in some way). Being nothing but a peculiar summation of these e^x functions, we might imagine the nature of a sinusoid will be relatively simple...and indeed it is (but perhaps not so simple as you might think). Nonetheless, the *nature* of these sinusoids remained beyond our reach for most of our recorded history, and even today relatively few understand their profound nature. Now, as we said, these sinusoids are only a simple summation of the *natural growth* function e^x, and as such they can't be all that difficult. Let's take a look.

MACLAURIN'S SERIES

We noted in the first book that we can approximate functions such as e^x, $\sin(x)$, $\cos(x)$, etc., using the Maclaurin series, which has the form:

$$f(x) = f(0) + f'(0)x + f''(0)x^2/2 + f'''(0)x^3/(2\cdot 3) + \ldots \quad (2.2)$$

where $f'(0)$, $f''(0)$, etc. are the 1st, 2nd, etc., derivatives, evaluated for an argument of zero. We will show how simple it is to use this series via three examples. The first is e^x—the function whose derivative is equal to itself. Now, anything raised to the zero power is equal to 1.0, so this makes Maclaurin's series for e^x very easy to evaluate—every derivative will yield a coefficient of 1.0 (when evaluated for an argument of zero) so:

$$e^x = 1 + x + x^2/2 + x^3/6 + x^4/24 + x^5/120 + \ldots + x^n/n! \quad (2.3)$$

You may prove this is correct by taking the derivative of this series of terms—the result will be the same as the original series—the derivative of this function is equal to itself.

The next function we approximate is the sine function. We know the value of a sine function is 0.0 at an argument of zero, and its derivative (a cosine function) is 1.0 for an argument of zero. The second derivative (a minus sine function) again yields 0.0 (for an argument of zero) and the third derivative (a minus cosine) is -1.0 (at zero), etc., etc. We therefore obtain a Maclaurin series for the sine function:

$$\sin(x) = x - x^3/6 + x^5/120 + \ldots \quad \text{--------} \quad (2.4)$$

Similarly the cosine function yields a series of:

$$\cos(x) = 1 - x^2/2 + x^4/24 + \ldots \quad \text{--------} \quad (2.5)$$

Note that the sine function contains all the odd powers of x while the cosine contains all the even powers. Most interestingly, e^x contains both the odd and even terms, which suggests we might somehow obtain the e^x function by summing the sine and cosine functions. In fact, we find these functions combine very neatly as follows:

$$e^{ix} = 1 + ix - x^2/2 - ix^3/6 + x^4/24 + ix^5/120 + \ldots \quad (2.6)$$

from which we may factor out the two series:

Nature of Sinusoids

$$\cos(x) = 1 - x^2/2 + x^4/24 + \ldots \quad \text{(2.7)}$$

and:

$$i\sin(x) = ix - ix^3/6 + ix^5/120 + \ldots \quad \text{(2.8)}$$

That is:

$$e^{ix} = \cos(x) + i\sin(x) \quad \text{(2.9)}$$

If we multiply this by A and substitute θ for x we get:

$$Ae^{i\theta} = A\cos(\theta) + i\,A\sin(\theta) \quad \text{(2.10)}$$

Now, from our development of sinusoids and Argand diagrams in Chapter 1, it's apparent the sine and cosine are the components of a unit length rotating vector (we will refer to rotating vectors as *phasors*), and it's further apparent we may write the components of *any* phasor as:

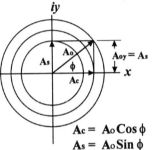

$A_c = A_o \cos \phi$
$A_s = A_o \sin \phi$

$$f(\theta) = A\cos(\theta) + i\,A\sin(\theta) \quad \text{(2.10A)}$$

We frequently write $f(\theta)$ as $A\,\underline{/\theta}$ but it's apparent $e^{i\theta}$ is a much better expression for a phasor (or vector). The i in front $A\sin(\theta)$ implies this component lies on the *imaginary* axis while $A\cos(\theta)$ lies on the *real* axis— we are dealing with the Argand mode of projection—the same we used in Fig. (1.1) on the first page to define the sine and cosine functions.

Note that the angle goes in the exponent of $e^{i\theta}$—in vector multiplication we will multiply the magnitudes and add the angles.

$$A_1 e^{i\theta_1} A_2 e^{i\theta_2} = A_1 A_2\, e^{i\theta_1} e^{i\theta_2} = A_1 A_2 e^{i(\theta_1 + \theta_2)} \quad \text{(2.11)}$$

Clearly, then, $e^{i\theta}$ is the mathematical expression for a phasor (or vector); however, we may repeat the above exercise for e raised to $-ix$:

$$e^{-ix} = \cos(x) - i\sin(x) \quad \text{(2.12)}$$

and by adding and subtracting (2.9) and (2.12) we may determine:

$$\cos(x) = (e^{ix} + e^{-ix})/2 \quad \text{(2.13)}$$

and:

$$\sin(x) = (e^{ix} - e^{-ix})/2i \quad \text{---------------} \quad (2.14)$$

So, what have we derived here? Back on the first page we lamented that simply writing $Sin(\theta)$ and $Cos(\theta)$ *hides* the mathematics that underlie these functions, and here we have *partially exposed* the underlying math...but what do we really have here? What do the expressions on the right side of the above equations really mean?

We know the imaginary exponents in the right hand side of (2.13) and (2.14) imply these are functions in the Argand plane—e^{ix} terms imply phasors. If we plot the right hand side of (2.13) on an Argand diagram we get Fig. 2.2 at right. From this it's apparent the *imaginary* components of these phasors will always cancel while the real components add[1] to twice the cosine of ϕ. As ϕ varies from 0 to 2π the cosine of ϕ is traced out on the real axis. As we know, we may then express this

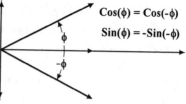

Fig. 2.2 - Complex Conjugates $e^{i\phi} + e^{-i\phi}$

cosine function *parametrically* [as a function of the parameter ϕ (or time)] in the usual Cartesian coordinate plane. It's apparent we will likewise get a projection of the sine function on the *imaginary* axis if we subtract the two exponential terms (see Eqn. 2.14).

2.1 THE "SELFISH" NATURE OF SINUSOIDS

The exponential expressions for sinusoids derived above expose much about the nature of these functions. We noted that e^x was a most unique function (being the function whose derivative equals itself), and here we see how sinusoids are constructed from these functions; so, do the derivatives of sinusoids really have similar characteristics.

In fact, direct examination of the sine function reveals this expected relationship. We know the vertical projection of a unit radius generates a sinusoid as it rotates, and it's also apparent any length of arc along the circumference of *a unit radius circle* must be equal to the

---

[1] An expression containing two terms is a "binomial." Two binomials of the form $a + b$ and $a - b$ are "conjugates." Complex numbers of the form $a + ib$ and $a - ib$ go by the impressive moniker **"complex conjugates."** Addition would yield $2a$ and multiplication would yield $a^2 + b^2$ ∴ $(Cos\phi + iSin\phi)(Cos\phi - iSin\phi) = 1$.

Nature of Sinusoids

included angle in radians. Now, the slope of a function is the *rise over the run*, but with sinusoids the *run* (i.e., *change* of the independent variable) is the arc encompassed (see Fig. 2.3).

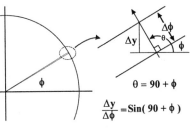

Fig. 2.3 - The Slope of a Sinusoid

We note the *displacement* of the tip of the radius is always at right angles to itself, so the direction of motion is equal to the angle of the radius plus 90° (i.e., $\phi + 90°$). From Figure 2.3 it's apparent that:

$$\Delta y / \Delta \phi = \text{Sin}(\phi + 90°) \qquad (2.15)$$

[Note: it's apparent that the hypotenuse of the little triangle (on which we base our equation for slope) in Fig 2.3 is equivalent to a radius that has rotated past 90°. The vertical projection of a radius is the sine of the angle ϕ, but $\Delta y/\Delta \phi$ is the *change in the sine function* divided by the *change in the independent variable*—it's the slope of the sinusoid. Furthermore, as is apparent from Figs. 1.2 and 1.3, the sine of (ϕ + 90°) is a cosine function. The slope, or derivative, of a sine wave is just a cosine wave.

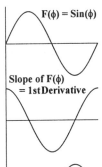

$$\Delta y / \Delta \phi = \text{slope } [\text{Sin }(\phi)] = \text{Sin}(\phi + 90°)$$
$$= \text{Cos }(\phi) \qquad (2.16)$$

Clearly, Figures 1.2 and 1.3 are the same thing except for the 90° displacement—we are only looking at different segments of the same curve. Just as clearly, then, if the derivative of a sine wave is just another sinusoid displaced by 90°, the derivative of a cosine wave must also be a sinusoid displaced by 90° (i.e., a minus sine wave). The 3rd derivative of a sine wave must yield a minus cosine and the 4th brings us back to our original sine function.

This is the relationship we wanted to expose: if e^x is the function whose derivative equals itself, then sinusoids (since they are composed of e^x functions) exhibit a similar characteristic in that their 2nd derivatives equal themselves (multiplied by -1); furthermore, their 4th derivatives are completely identical to themselves. This relationship is more clearly seen if we simply take the derivatives of Eqns. (2.13) and (2.14):

$$d\cos(x)/dx = \tfrac{1}{2}\, d(e^{ix} + e^{-ix})/dx = \tfrac{1}{2}[i(e^{ix} - e^{-ix})]$$
$$= -(e^{ix} - e^{-ix})/2i \qquad \text{------------------} \quad (2.17)$$

and:

$$d\sin(x)/dx = \tfrac{1}{2i}\, d(e^{ix} - e^{-ix})/dx = i(e^{ix} - e^{-ix})/2i$$
$$= (e^{ix} + e^{-ix})/2 \qquad \text{------------------} \quad (2.18)$$

2.2 THE UNIVERSALITY OF SINUSOIDS

Just as e^x (i.e., the law of natural growth) describes very fundamental relationships in our world, its offspring (i.e., sinusoids) are likewise profound. In mechanical engineering, for example, there are three fundamental phenomena in physically real linear systems (i.e., *inertial mass*, *spring force*, and *energy loss mechanisms*.

Now, these three *elements* do completely different things. Friction, for example, being the pre-eminent example of energy loss, *attenuates* a "signal" (e.g., the torque applied to a shaft supported by bushings will be reduced by the friction of these bearings). This phenomenon is represented mathematically as a *multiplied coefficient*.

The other two elements (inertia and springs) are, as we said, completely different—we can illustrate the *inertia/spring* functions by clamping the end of a ruler to the edge of a desk (with the palm of one hand) and "twanging" the suspended end (the thumb of your other hand serving as twanger). The ruler vibrates in a sinusoidal motion at its *eigenfrequenz*. We all know the reason this happens is because the energy continues to be transferred from the spring element to the inertial element and back (air resistance providing the dominant energy loss mechanism in this case). We note the *driving function* for this experiment is the twang given by your thumb (even without a lot of detailed analysis it's apparent something more interesting than bearing friction is going on here).

What *is* going on here? Well, according to Sir Isaac, if we apply a force to an unrestricted mass it will not just move—it will *accelerate*. It will accelerate in the *direction of the force* regardless of whether the force is directed with, against, or perpendicular to the present direction of motion. It's implicit that if there's no force, there's no acceleration—and vice-versa. Recognizing that acceleration is the rate of change of velocity,

Nature of Sinusoids

we state the above mathematically as:

$$F = mA = m\, dv/dt \qquad (2.19)$$

where:
- F = force
- m = mass
- A = acceleration
- dv/dt = rate of change of velocity = A

It was Robert Hook, however, who told us (even before Sir Isaac revealed the nature of inertia) that when we stretch a spring, the elongation is proportional to how hard we pull. That is, the distance the ruler bends is proportional to the force applied by your thumb. Mathematically, this is another very simple relationship:

$$F = K\Delta x \qquad (2.20)$$

where:
- K = spring constant
- Δx = increment spring has stretched

We have a small problem here—the equation for force on a spring is expressed in terms of position (or distance), while the equation for force on a mass is written for acceleration or rate of change of velocity. This presents no real problem of course, for just as acceleration is the rate of change of velocity, we know velocity is the rate of change of position vs. time. It's therefore apparent that acceleration is the 2nd derivative of *position* vs. time:

$$F = m\, d_2 x/dt^2 = mx'' \qquad (2.21)$$

where:
- $d_2 x/dt^2$ = 2nd derivative of position vs. time
- $x'' = d_2 x/dt^2$

Okay, now we can compare oranges to oranges, but since mass involves the 2nd derivative it's a little like comparing Florida oranges to "some other place's" oranges. Now, ol' Newt also told us that for every *action* there's always an equal and opposite action. Clearly, in any closed system (such as our twanged ruler), the sum of all the forces at any time must be equal to zero (i.e., force is the action we speak of and, when summed (so long as we account for *all* of the forces), this equal and opposite characteristic must yield zero). If we ignore the force of the energy losses in our twanged ruler example we can write the equation:

$$\textit{accelerating force - spring force} = 0$$
$$mx'' - K\Delta x = 0 \qquad \text{-------} \qquad (2.22)$$

This simple little equation tells us the following: As we bend the end of the ruler via force from our thumb, the molecules of the ruler are pulled from their equilibrium position creating an opposing *restoring force* (equal and opposite). When we slip our thumb over the edge, suddenly all of the restoring force is applied to acceleration of the mass (of the ruler) according to Eqn. (2.22). The ruler accelerates; which, in time, generates a velocity; which, in time, moves the ruler toward the position of equilibrium (i.e., where $\Delta x = 0$). As the ruler moves toward the position $\Delta x = 0$ the spring force diminishes [Eqn. (2.20)], and consequently the acceleration decreases [Eqn. (2.19)] until, at $\Delta x = 0$, there is no force anywhere in this system (remember, we are ignoring wind resistance).

Now, the ruler's mass is still in motion and, unless some force acts on that mass, it will *continue* in that motion; however, the motion of the ruler immediately begins to wind the spring in the opposite direction, and force is (again) generated. This restoring force acts to accelerate the mass of the ruler in the opposite direction (i.e., slowing the velocity). As the ruler continues to deflect and the spring force increases, the negative acceleration (being proportional to the force) must increase, and velocity is quickly brought to zero. Clearly, the deflection of the ruler is at its greatest (there being no velocity to carry it farther), so the restoring spring force is at its greatest (negative), which means the acceleration is at its greatest—only velocity is zero. From here we go through the identical scenario in the opposite direction....all this from Equation (2.22).[2]

But the cause for celebration doesn't end here—in electrical engineering inductance is analogous to inertial mass—a capacitor mimics a spring—and resistance is analogous to friction. Consequently, if we replace m in Eqn. (2.22) with L (the term for electrical inductance), and K with $1/C$ (where C is the symbol for electrical capacitance) we get the most fundamental equation for an electrical circuit response. Similar conceptual elements are found in fluid dynamics, acoustics...in fact, this general equation is found throughout engineering. Unfailingly we come up with what is called a *second order, linear, differential equation*—Eqn. (2.22).

---

[2] Physicists sincerely search for an *equation of everything*, and we sincerely wish them well; but, here, in this model of mechanical motion (if we only include the element of friction) we very nearly have a *mechanical motion equation for everything*.

Nature of Sinusoids

2.3 EIGENFUNCTIONS

The word *eigen* (pronounced *eye'-gun*) comes from the German language and is translated as "own", "peculiar," "self," or "specific." German words, however, are properly defined in German (just as English words are defined in English), and a direct translation of words is greatly aided by examples of common usage. The word "Nutz" (pronounced *noots*), for example, means *usefulness*; so "eigen-nutz" denotes the notion we refer to as *self-interest*. Even more enlightening, "eigen-nutz*ig*" means "selfish." "Eigen-frequenz" denotes the phenomenon American engineers refer to as *natural frequency*. The German word "Lob" (long *o* as in *low*) means *praise*, and "Eigen-lob" is, of course, *self-praise*.

What, then, is an *eigen-function*?

2.4 STURM-LIOUVILLE

The general applicability of trigonometric series was something Fourier hadn't proven when he used this technique to solve his famous problem in *heat flow* in the *early*-nineteenth century. By *mid*-nineteenth century things had progressed and two mathematicians (J.C.F. Sturm and Joseph Liouville)[3] worked on the general solution of these second order ordinary differential equations (equations of this form are now called Sturm-Liouville equations). If, in Equation (2.22), we move the $Kf(x)$ expression to the right-hand side:

$$Mf'''(x) = Kf(x) \qquad (2.22A)$$
$$\text{then:} \quad Mf'''(x)/K = f(x) \qquad (2.22B)$$

where the constants M and K are not necessarily mass and springs. In this form we see (since M and K are constants), the second derivative of $f(x)$ is equal to itself (modified at most by a multiplied constant)!

Now, we know a function whose 2nd derivative is equal to itself multiplied by a constant. Not only does the second derivative of a *sine wave* equal itself (multiplied by a constant -1) but so does the 2nd derivative of a *cosine wave*. On successive differentiations, these functions *are self-generating!* You may begin to see where all this is going, but let's

---

[3] In Russia, this work is attributed to the *prominent* Russian mathematician, V.A. Steklov.

consider another relationship apparent in our development of Eqn. (2.22). In the preceding illustration, the *driving* function is the twang from your thumb, and the *response* function is the oscillatory motion of the ruler. We note specifically that the *response function* doesn't look like the *driving function*. That is, the *twang* we give the ruler is (more or less) an impulse; but the response is (more or less) a sinusoid. This is, in fact, more often the case than not—we don't expect the response to look like the driving function—the response is frequently a sinusoid regardless of the shape of the driving function.

So, what happens if we *drive* the system with a sinusoid? The only possible operations a linear system can perform on the input function are attenuation, differentiation, and integration. We also know that sinusoids reproduce themselves on differentiation and integration (as opposed to, say, square waves which yield completely different functions via these operations). Since attenuation only scales a function, if we drive a linear system with a sinusoid, we will get a sinusoid out. The most that can happen is that we shift the phase or change the amplitude, but if we put a sinusoid in we get a sinusoid out! The sinusoid is an *eigenfunction* of a linear system.

"These *eigenfunctions*," says Sturm (in a heavy Swiss/German accent), "Herr Fourier can ein Trrrrransformer mit build!"

"Mais oui! But of course," says Liouville (in a light Swiss/French accent). "Yet only when bound-up by our *conditions*—n'est-ce pas?"

"Conditions? What conditions?" [we ask with a blank expression].

Why *boundary conditions*? Consider this—we know another function whose 2nd derivative is equal to itself (as are all of its derivatives): e^x. But e^x is *not* an eigenfunction of linear systems because it doesn't satisfy the boundary conditions. Specifically, *the boundary conditions demand that the function (or its derivatives) equal zero at the end points (i.e., its boundaries)*. Sine and cosine waves meet these conditions—real exponentials do not (ever get the feeling these guys are playing with loaded dice?) We know that a sine wave begins with a value of zero, and if we scale the argument properly (i.e., select the right frequency), we know we can make it equal zero at the end point too. If we do *this* we also know that any integer multiple of this particular scaling factor (i.e., harmonics of this fundamental) will also begin and end with zero—and likewise qualify as eigenfunctions. Furthermore, since the

Nature of Sinusoids 17

derivatives of *cosine waves* will be zero at their beginning and ending points, they too qualify as eigenfunctions. What about e^x? Sorry....

Our primary interest in *Sturm-Liouville* is exposing the relevance of eigenfunctions to real world, linear, time invariant (e.g., spring/mass/friction) systems. Understand that the sinusoid really is an *eigenfunction* of real world *linear* systems. If we drive our circuit with a sinusoid (at a given frequency), we will get a sinusoid out (at the same frequency). When we twang a ruler it really *does* oscillate in a sinusoidal motion. On the other hand, it's apparent to anyone who has ever run a square wave through a low pass filter that *square waves are not eigenfunctions of linear systems*—what you get out is *not* what you put in. For that matter, neither are the other wave shapes that are grossly different from sinusoids.

So, why do Sturm and Liouville state their demands in such a round-about way? Simply because eigenfunctions may be any variation on a sinusoid that meets the Sturm-Liouville boundary conditions! For example, the Laplace transform uses an exponentially decaying sinusoid described as $e^{-(\rho+i\beta)}$ (you will recognize that this is a sinusoid multiplied a real exponential function $e^{-\rho} \cdot e^{-i\beta}$). This is not the only variation that qualifies as an eigenfunction, of course, and we shall discover others as we progress through this book.

We begin to see a little of the *nature* of these sinusoids, but there is considerably more. When we work with them mathematically we discover they exhibit very unique characteristics, and we will only understand their true nature when we begin to understand these mathematically unique characteristics.

CHAPTER 3

THE MATHEMATICALLY UNIQUE SINUSOID

If we have sinusoids exactly *in phase*, we can obviously add them by simply summing their amplitude coefficients; but, clearly, we *can't* simply add sinusoids when they have different phases (if the phase equals

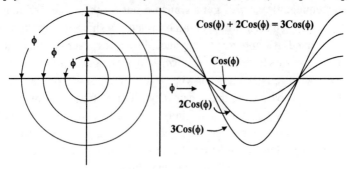

Figure 3.1 - *In Phase* Sinusoids

±180° addition becomes subtraction—see Fig. 3.4).

We can break a phasor into sine and cosine components, and it's apparent that, if the phasor rotates, the sinusoid generated is equal to the sum of the sinusoids generated by the phasor components[1]. That is, the

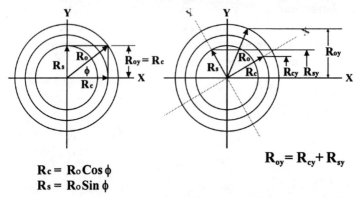

$R_c = R_o \cos \phi$
$R_s = R_o \sin \phi$

$R_{oy} = R_{cy} + R_{sy}$

Figure 3.2 - Rotating Phasor Components

---

[1] As a phasor's components rotate they too may be broken into components. Clearly the summation of a phasor's components must yield the original phasor regardless of how these are rotated; therefore, the sum of *y* axis projections of the components must equal the *y* axis projection of the phasor at all rotations.

The Unique Sinusoid

components of static phasors will lie along the horizontal and vertical axes and therefore we may add the phasor *components* (which lie on the same axis). We may thus add sinusoids of arbitrary phase by *complex addition* (i.e., add imaginary components to imaginary, and real with real).

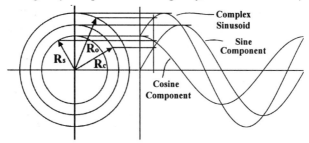

Figure 3.3 - Complex Sinusoids

3.1 SUMMATION OF *UNEQUAL FREQUENCY* SINUSOIDS

The above scheme for addition works well when we deal with sinusoids of the same frequency; however, it will obviously not work when the sinusoids to be added are of different frequency (since the phase will constantly change). This is illustrated in Fig. 3.4—note that the value of the *summation* constantly changes as we progress along the wave.

This remarkable result is known as "beat frequency modulation." As the phase between the two sinusoids slowly *walks*, the amplitudes first add, then subtract, then add again. Having completed 360° of phase shift they can only repeat this cycle. Note that the resultant waveform is still essentially a sinusoid, albeit an *amplitude modulated* sinusoid. The term "beat frequency" comes from the distinctive *beating* sound of two musical tones which are not quite in tune. The beating effect is obvious, but a more subtle, unique characteristic of sinusoids is also working here. Note that the

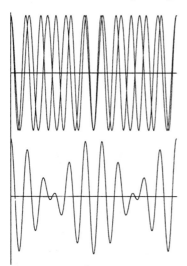

Figure 3.4 - Summation of Sinusoids

summation of two cosine waves, for example, equals:

$$f(t) = Cos(2\pi f_1 t) + Cos(2\pi f_2 t) \quad \text{------------} \quad (3.1)$$

These two sinusoids start *in-phase* but immediately begin walking *out-of-phase*. The next time they will be back in-phase will be when the higher frequency sinusoid has completed *one full cycle more* than the lower frequency sinusoid, but this equation becomes much more interesting when we consider its trigonometric identity:

$$Cos(A) + Cos(B) = 2[Cos(\tfrac{1}{2}(A+B))Cos(\tfrac{1}{2}(A-B))] \quad (3.2A)$$

substituting the terms used in (3.1):

$$Cos(2\pi f_1 t) + Cos(2\pi f_2 t) = 2[Cos(2\pi t \tfrac{f_1+f_2}{2})Cos(2\pi t \tfrac{f_1-f_2}{2})] \quad (3.2)$$

Let's hold off discussing this equation for a moment...except to note that the result of summing two sinusoids of different frequency is equivalent to *multiplying* two sinusoids of suitably selected frequencies. Surely this result is unique to sinusoids...but let's not get ahead of ourselves....

Let's take a candid look at what happens when we *multiply* two sinusoids. If we multiply one sinusoid (of high frequency) by a second (of lower frequency) we must obtain the pattern shown at right. It is more or less obvious that, when we multiply two sinusoids together in this manner, the "envelope" of the higher frequency sinusoid will take the shape of the lower. The most interesting thing here, of course, is that the resulting modulation pattern is quite similar to the envelope obtained in Fig. 3.4. In fact, the modulated waveform shown here is *identical* to the waveform obtained by summing those sinusoids. That is, modulation (i.e., multiplication of one

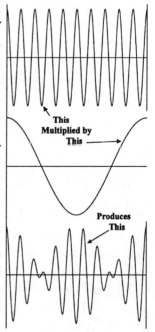

Figure 3.5 - Modulation

The Unique Sinusoid

sinusoid by another) produces what we call "sidebands" about the modulated signal.

We may show this side-band relationship of modulated signals by employing another well known trigonometric identity:

$$Cos(A)Cos(B) = \frac{Cos(A+B)+Cos(A-B)}{2} \quad (3.3A)$$

If $A = 2\pi f_c$, and $B = 2\pi(0.1f_c)$ then the above becomes:

$$Cos(2\pi f_c)Cos(2\pi(0.1f_c)) = \frac{Cos(2\pi 1.1f_c)+Cos(2\pi 0.9f_c)}{2} \quad (3.3)$$

Eqns. (3.2) and (3.3) show the same thing. When we multiply sinusoids, two new side band sinusoids are produced *(note that the original sinusoids disappear and the two new sidebands are of different frequency than either of the sinusoids multiplied together)*.

This unique aspect of sinusoids has profound significance. This curious result (i.e., of sinusoid multiplication) is the underlying mechanism of *convolution* (which plays such a prominent role in understanding and applying Fourier analysis). It will be worth digging a little deeper here.

3.2 SINUSOID MULTIPLICATION

Back on page 11 we noted multiplication of complex exponentials (i.e., vector or phasor multiplication) required multiplication of the magnitudes and summation of the angles. On page 12 we gave the geometric interpretation of Equations (2.13) and (2.14) which generate pure sinusoids. Let's now look at a similar geometric interpretation of phasor multiplication.

Multiplication is only a process of repeated addition. If we multiply a vector (of magnitude 2) lying along the X axis by another vector (of magnitude 3) also on the X axis, the resultant is obtained by summing 2 into an "accumulator" 3 times. This is precisely the operation we perform in elementary school multiplication, but consider the example of a vector of magnitude 3 on the X axis and another of magnitude 2 *on the Y axis*. We must now multiply

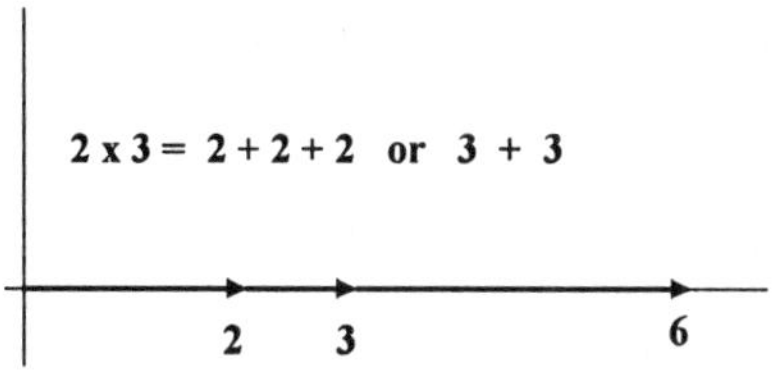

Figure 3.6

a quantity of 2 *with π/2 radians rotation* by a quantity of 3. Now, one of the numbers is being added to itself and the other tells how many times to sum it into the *accumulator*. According to our development of complex addition, we would take our 2 on the Y axis and sum it into the *pot* three times yielding 6 *on the Y axis*. It would be silly to think the *angle* should be multiplied by 3.

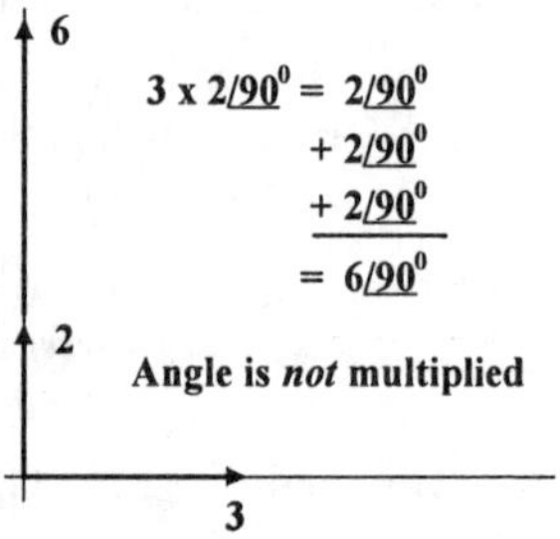

$3 \times 2\underline{/90}^0 = 2\underline{/90}^0$
$+ 2\underline{/90}^0$
$+ 2\underline{/90}^0$
$= 6\underline{/90}^0$

Angle is *not* multiplied

Figure 3.7

If we perform this multiplication the other way around and sum 3 *on the X axis* into the pot 2 *with π/2 radians rotation* times, then before we throw 3 *on the X axis* into the pot, we must rotate it by π/2 radians. The thing we are throwing into the pot then is 3 at π/2 radians—the only way we can actually sum quantities is if they are aligned (in the same direction). So, first we take 3 /0, rotate it by π/2, throw it into the pot; then we take 3 /0 again, rotate it by π/2, and *sum it in*—yielding 6 at π/2.

Clearly, then, if we have 3 on the *Y axis* multiplied by *2 on the Y axis* we must sum 3 at 90° into the accumulator 2 at 90° times. We must rotate 3 at 90° by the angle of the multiplier, and sum it into the pot the number of times indicated. If we do it this way the complex quantities being summed into the pot will always be at the same angle—as they must.

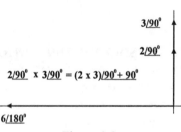

$2\underline{/90}^0 \times 3\underline{/90}^0 = (2 \times 3)\underline{/90}^0 + 90^0$

Figure 3.8

Having reviewed this very basic mechanism, we're now ready to look at the multiplication of sinusoids (as these were represented back on page 10). Figure 2.1 allows us to visualize the solitary cosine function in terms of a simple pair of phasors—we need not invoke thoughts of series expansions, etc. [Understand here that we *must* use this representation of a cosine function—we *must* use this *pair* of complex conjugate phasors if we hope to see what is really happening mathematically when we multiply two sinusoids.]

So, then, what happens (geometrically) when we multiply two of these *geometrical* cosine functions?

The Unique Sinusoid

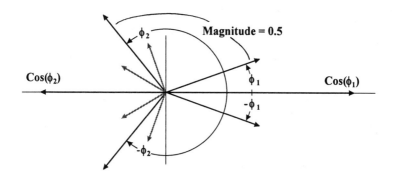

Figure 3.9 - Multiplication of Two Cosines

Each of these cosines has a pair of complex conjugate phasors, and if we want to multiply these cosine functions, we must multiply one pair of complex conjugates by the other pair—we must multiply each phasor of one pair by both phasors of the other. As we have just seen, when we multiply phasors we multiply their magnitudes and add their angles; so, if we stop reading and try that scheme on the figure above, it's apparent we will obtain the following: since the magnitudes of all the phasors to be multiplied are 0.5, the magnitude of all the products will be 0.5×0.5 = 0.25. Furthermore, since we are dealing with complex conjugate pairs (i.e., the angles of the pair belonging to either cosine are the negatives of each other), the angles of the products will be the sums and differences of the multiplied phasors' angles (see Fig. 3.9). We will obtain two *new* complex conjugate pairs of phasors (i.e., two new cosine functions) which are superimposed (i.e., summed together—Fig 3.9). Note carefully that, if one angle is large and the other small, the two new phasors will *group themselves* about the larger angle (i.e., they must be spaced equally ahead of and behind the larger angle by + and - the smaller angle).

Okay, now we replace the static angle ϕ by our kinetic representation of an angle ωt (i.e., $2\pi ft$) so that our phasors rotate. If the two frequencies are different, and we multiply our two cosine functions point by point as the phasors rotate, we will be performing the operation we described above as *modulation*. Again, note that the faster moving phasor (i.e., higher frequency of rotation) will immediately obtain a larger angle, and the products of the multiplication will group themselves about this faster moving phasor. The frequency of rotation of one complex conjugate pair will be equal to the higher frequency *plus* the lower—the other will be the higher *minus* the lower. Mathematically:

$$Cos(2\pi f_o t)Cos(2\pi f_1 t)) = \frac{Cos(2\pi f_o t + 2\pi f_1 t) + Cos(2\pi f_o t - 2\pi f_1 t)}{2} \quad (3.4A)$$

$$Cos(2\pi f_o t)Cos(2\pi f_1 t)) = \frac{Cos(2\pi (f_0 + f_1)t) + Cos(2\pi (f_o - f_1)t)}{2} \quad (3.4)$$

or, in exponential form:

$$= \tfrac{1}{4}[e^{i2\pi(fo+f1)t} + e^{-i2\pi(fo+f1)t}] + \tfrac{1}{4}[e^{i2\pi(fo-f1)t} + e^{-i2\pi(fo-f1)t}] \quad (3.5)$$

where f_0 is the higher frequency

These new sinusoids (i.e., the products of two other sinusoids) are, as we said, referred to as *sidebands* since, in the frequency domain, they will stand on either side of the higher frequency sinusoid. As we also said, this mechanism is the basis of convolution, and Fourier analysis itself... as we will now explain.

3.3 THE FOURIER TRANSFORM

When we simply *accept* sines and cosines as presented in Chapter 1, we *short-circuit* everything we have covered in Chapters 2 and 3. It's only when we include the concepts presented above that we obtain a mathematically consistent understanding of sinusoids. Now that we understand what really happens we may use the derived equations to examine Fourier's transform and it's application.

First, let's look at multiplication of two *real* sinusoids of arbitrary phase (but same frequency):

$$Cos(\omega t + \phi_1)Cos(\omega t + \phi_2) =$$
$$[(e^{i(\omega t + \phi 1)} + e^{-i(\omega t + \phi 1)})(e^{i(\omega t + \phi 2)} + e^{-i(\omega t + \phi 2)})]/4 \quad (3.6)$$

[We have omitted magnitudes here since they always *factor out* and contribute nothing to the illustration.] You may easily show that the right-hand side of (3.6) reduces to:

$$[(e^{i(2\omega t + \phi 1 + \phi 2)} + e^{-i(2\omega t + \phi 1 + \phi 2)}]/4 + [e^{i(\phi 1 - \phi 2)} + e^{i(\phi 2 - \phi 1)}]/4 \quad (3.7)$$

This is truly a magnificent expression—the two exponential terms on the left are just a sinusoid (a half amplitude, phase-shifted cosine wave). *When*

The Unique Sinusoid

we multiply two sinusoids of arbitrary phase, we get another sinusoid, the frequency of which is apparently doubled with the phase shifted by the sum of the phase angles.

In addition to this sinusoid, there are two terms in the right-hand set of brackets of Eqn. (3.7). These concern the phase of the sinusoids:

$$e^{i(\phi_1-\phi_2)} + e^{i(\phi_2-\phi_1)} = [\cos(\phi_1-\phi_2) + i\sin(\phi_1-\phi_2)] + [\cos(\phi_2-\phi_1) + i\sin(\phi_2-\phi_1)]$$

Since $\sin(\phi_2-\phi_1) = -\sin(\phi_1-\phi_2)$, and $\cos(\phi_1-\phi_2) = \cos(\phi_2-\phi_1)$ we have:

$$e^{i(\phi_1-\phi_2)} + e^{i(\phi_2-\phi_1)} = [2\cos(\phi_1-\phi_2)] \quad\cdots\cdots\cdots\cdots \quad (3.8)$$

These are unit vectors at plus and minus the *difference* between the phase angles and are thus *complex conjugates* (i.e., they generate equal and opposite imaginary components which cancel—see footnote p. 10).

This relationship may be more apparent from Fig. 2.1 (p. 10). This cosine term [i.e., Eqn.(3.8)] is, in Eqn.(3.7), an added constant proportional to the cosine of the *phase difference*. When the two sinusoids are in phase this will be a maximum of 2.0—when they are 90° out of phase the added constant is zero. Now, when we take the *average* of the complete expression (i.e., Eqn. 3.7) over 2π radians, any *sinusoid* will contribute zero to the average value, and the added constant [i.e., Eqn.(3.8)] will be the only thing remaining. *It is therefore apparent that when we multiply a waveform (composed of a single complex sinusoid) by a cosine wave of the same frequency, the <u>cosine component</u> in the complex waveform will yield an average value—the <u>sine term</u> will yield none. If we multiply this complex waveform by a sine wave, the <u>sine component</u> of the waveform will yield an average and the cosine will yield none.* Equation (3.7) reveals everything—frequency doubling, phase angles adding, amplitude divided by two...even orthogonality. A magnificent expression.

We must consider one last technicality here—the multiplication of a sinusoid by the complex exponential:

$$\{(e^{i(\omega t + \phi_1)} + e^{-i(\omega t + \phi_1)})/2\} \; e^{i(\omega t + \phi_2)} = ?? \quad\cdots\cdots \quad (3.9)$$

This is an important exercise, for it represents the most common form in which the Fourier Integral is written:

$$F(\omega) = \int_{-\infty}^{+\infty} f(t) \, e^{-i\omega t} \, dt \quad\cdots\cdots\cdots \quad (3.10)$$

where $f(t)$ in (3.10) may be represented by the expression for a single complex sinusoid as in (3.9). For the Fourier transform, the phase of the complex exponential (i.e., ϕ_2) is zero, so we may drop this term and expand Eqn. (3.9) as follows:

$$\{(e^{i(\omega t + \phi 1)} + e^{-i(\omega t + \phi 1)})/2\} \, e^{i\omega t} = (e^{i(2\omega t + \phi 1)} + e^{-i(\omega t)} \cdot e^{i(\omega t)} \cdot e^{-i\phi 1})/2$$

and as we explained in the footnote on page 10, $e^{-i(\omega t)} \cdot e^{i(\omega t)}$ indicates a product of complex conjugates and is simply equal to 1. Therefore:

$$\{(e^{i(\omega t + \phi 1)} + e^{-i(\omega t + \phi 1)})/2\} \, e^{i\omega t} = (e^{i(2\omega t + \phi 1)} + e^{-i\phi 1})/2 \quad \text{-----} \quad (3.11)$$

so multiplication of a real sinusoid by the complex exponential $e^{i(\omega t)}$ results in a *phasor* which rotates at $2\omega t$ and has a phase angle equal to ϕ_1. Now, there is also a vector constant $e^{-i\phi 1}/2$ which, we recognize, breaks into the components $[\cos(\phi_1) - i\sin(\phi_1)]/2$. If the sinusoid in $f(t)$ has magnitude M_n, this operator breaks it into real and imaginary phasor components (divided by 2).

The integral finds the area under the curve described by the right-hand side of Eqn. (3.11), and the oscillating nature of the first complex exponential will contribute nothing to the final value. Integration of Equation (3.11) then, must ultimately yield only the *phasor* of the complex sinusoid of $f(\omega)$. That, of course, is precisely what we expect from the Fourier transform.

3.4 COMPONENT ORTHOGONALITY

Orthogonality is the phenomenon of the product of two sinusoidal functions yielding an average value of *zero* (averaged over any number of full cycles) when either:

1) the sinusoids are de-phased by 90° or,
2) the sinusoids are of integer multiple frequencies.

This condition of orthogonality is imperative if we are to resolve functions into their frequency domain components by the mechanism of the DFT. We have just shown that, when we multiply two sinusoids *of the same frequency*, using the exponential expressions we have derived for sinusoids, orthogonality for 90° de-phased sinusoids is obtained. We must now expand that demonstration to show orthogonality for integer multiple

The Unique Sinusoid   27

frequency sinusoids. Before we undertake this demonstration, however, we must establish some fundamental facts. For example, the average value of a sinusoid (taken over a full cycle) is *zero*. If you do not see the validity of this, you are assigned the homework of proving (see Appendix 1.1) that the average value of a sinusoid is zero. If you prove otherwise, it will not be necessary to complete the rest of this chapter...you're way ahead of the rest of us....

Having established that its average value is zero, it's reasonably apparent that adding a full cycle of sinusoid to some function will not change the *average value* of that function (you may have already foreseen this profound observation but, again, if not...homework).

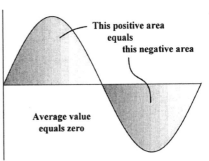

Fig 3.10 - Average Value of Sin(x)

Finally, before we tackle different frequency sinusoids, let's review orthogonality of same frequency sinusoids. We have illustrated multiplication of same frequency sinusoids by considering their exponential form:

$$\cos(\omega t + \phi_1)\cos(\omega t + \phi_2) =$$
$$[(e^{i(\omega t + \phi_1)} + e^{-i(\omega t + \phi_1)})(e^{i(\omega t + \phi_2)} + e^{-i(\omega t + \phi_2)})]/4 \quad (3.6)$$

and as we pointed out, the right-hand side of (3.6) reduces to:

$$[(e^{i(2\omega t + \phi_1 + \phi_2)} + e^{-i(2\omega t + \phi_1 + \phi_2)} + e^{i(\phi_1 - \phi_2)} + e^{i(\phi_2 - \phi_1)})]/4 \quad (3.7)$$

Now, we immediately recognize that the left-most two terms of expression (3.7) are nothing but a sinusoid:

$$[(e^{i(2\omega t + \phi_1 + \phi_2)} + e^{-i(2\omega t + \phi_1 + \phi_2)})]/4 = \tfrac{1}{2}\cos(2\omega t + \phi_T) \quad (3.12)$$

where $\phi_T = \phi_1 + \phi_2$

and the right-most two terms yield a constant determined by the phase difference between the two sinusoids:

$$[\cos(\phi_1 - \phi_2) + \cos(\phi_2 - \phi_1)]/4 = \cos(\phi_1 - \phi_2)/2 \quad (3.13)$$

From our introductory remarks, the sinusoid [Eqn. (3.12)] contributes nothing to the average value of this function, but the average value of a constant term is just the constant term. So, then, the average value of the product of two sinusoids will depend on the phase relationship between the two sinusoids—if the two are *in phase* (i.e., $\phi_1-\phi_2 = 0$) we will obtain an average value equal to one-half the product of the amplitudes of the two sinusoids! If, however, the phase relationship is $\pi/2$ (i.e., $\phi_1-\phi_2 = 90°$) the average value of the product will be *zero!* This latter result is the phenomenon of *orthogonality between sine and cosine functions*!

So, what about orthogonality between sinusoids of different frequencies then (which is the real question here)? Returning to our consideration of the average value of a sinusoid, we reiterate that it is necessary to deal with *full cycles* of a sinusoid for the average value to be zero. We can always satisfy this requirement by scaling the frequency term so that exactly one full sinusoid fits into the domain of the function we are working with. That is, if our sinusoid is given by $Cos(2\pi f_0 t)$, and the domain is $t = 0$ to T_0, we simply make $f_0 = 1/T_0$. When $t = T_0$ then, and we are at the end of the domain, the value of the argument will equal 2π and we will have constructed one full cycle of sinusoid (within the domain of definition). It's also apparent that if we limit frequencies to integer multiples of f_0, then every allowed frequency will provide an integer number of full cycles within the domain, and every sinusoid at an allowed frequency will have an *average value* of *zero* (see Fig. 3.9).

Allowing only integer multiple frequencies then, what happens when we multiply two sinusoids? Equation (3.38) must become:

$$Cos(\omega t + \phi_1)Cos(n\omega t + \phi_2) =$$
$$[(e^{i(\omega t + \phi 1)} + e^{-i(\omega t + \phi 1)})(e^{i(n\omega t + \phi 2)} + e^{-i(n\omega t + \phi 2)})]/4 \qquad (3.14)$$

where **n** takes on integer values only.

We will simplify the situation here by making both ϕ_1 and ϕ_2 zero so they disappear from the equation. The reason for this will be obvious at the end of this section. Now when we expand the expression on the right side of this equation we obtain:

$$[e^{i(\omega t + n\omega t)} + e^{-i(\omega t + n\omega t)} + e^{i(\omega t - n\omega t)} + e^{-i(\omega t - n\omega t)}]/4 \qquad (3.15)$$

Factoring ωt and collecting terms in the exponentials results in:

$$[e^{i[\omega t(1 + n)]} + e^{-i[\omega t(1 + n)]} + e^{i[\omega t(1-n)]} + e^{-i[\omega t(1-n)]}]/4 \qquad (3.16)$$

The Unique Sinusoid

There! It is done—it is finished—it is beautiful!

What? You don't see it? *Mein Gott im Himmel!* You recognize that the two exponentials on the left-most side are in the form of a cosine and the two on the right-most side are also of the cosine form. You also recognize that n is an integer multiple of the fundamental frequency. *Look at this equation!* When n equals 1.0 we are multiplying sinusoids of the same frequency. This equation then yields the following:

$$[(e^{i[2\omega t]} + e^{-i[2\omega t]} + e^{i[0]} + e^{-i[0]})]/4 \qquad (3.17)$$

Now, $e^{i0} + e^{-i0} = 2$ (see footnote page 10). Furthermore, the first two exponentials are a cosine of twice the frequency of the input functions—this is simply the equation for multiplying two sinusoids of the same frequency that we saw earlier—we already know this.

The *really* beautiful thing here is that this equation makes the result of multiplying sinusoids of different frequencies (i.e., when $n \geq 2$) obvious. What? You don't see that either? *Ma-ma mia!* When n equals 2, this equation becomes:

$$[(e^{i[3\omega t]} + e^{-i[3\omega t]} + e^{i[\omega t]} + e^{-i[\omega t]})]/4 \qquad (3.18)$$

Both the left-most pair of exponentials *and* the right-most pair are cosine functions! Both of these have an average value of zero! The average value of the product of a sinusoid multiplied by a second sinusoid of twice the frequency is *zero!*

Surely you see what happens when $n = 3$. No?? *Sacre bleu!*

$$[(e^{i[4\omega t]} + e^{-i[4\omega t]} + e^{i[2\omega t]} + e^{-i[2\omega t]})]/4 \qquad (3.19)$$

We still get two cosine waves. Only the frequencies are different. The product of two sinusoids with a 3:1 frequency ratio yields an average value of zero *again.*

By now you can probably see where all of this is going. You do?? *Oooooh wow!* You recognized the general relationship for the product of two sinusoids of different frequency (f1 and f2) is two other sinusoids whose frequencies equal (f1 + f2) and (f1 - f2)? You saw that only when f1 = f2 will the difference equal zero, and only a frequency of zero yields a constant term? That otherwise, we can keep increasing n until Arizona freezes over, and we will never again get an average value out of the product of two sinusoids of integer multiple frequencies? You see, then, that *cosines of different frequency must be orthogonal functions*.

Now, some of you will recall that we omitted the phase term from our equations at the beginning of this illustration, and you may suppose that we have *cheated* by a considerable amount. If thoughts such as these are floating through your head right now, cheer up—I have a real treat in store for you. Remember when I said the reason for omitting the phase terms would become obvious later? Weeell, the *mechanics* of going through this illustration *with* the phase terms is not significantly different from going through it *without* the phase terms—we have left the exercise of orthogonality vs. phase *and* frequency as "homework." It *will* count as part of your grade. Showing what happens to *sine* waves alone earns a C+.

Those who *do not* think we have cheated, however, and see the relationship between phase, frequency and orthogonality as intuitively obvious may continue on.

We still haven't completely exposed the nature of these sinusoids, but we will revisit this as we progress through the book. As we reveal the nature of these sinusoids, you will see this really *is* beautiful. There is a famous line from Sidney Lanier's "The Marshes of Glynn":

> "...Tolerant plains, that suffer the sea
> and the rains and the sun
>
> Ye spread and span
> like the catholic[2] man
> who hath mightily won
>
> God out of knowledge
> and good out of infinite pain
> and sight out of blindness
> and purity out of a stain.

When the pieces fit together so precisely we realize God is indeed in his heaven and the gears that make the world go 'round are greased and running smoothly. The problem is man's ignorance.

---

[2] Catholic in the sense of universal—broad in sympathies and interests.

CHAPTER 4

NETWORK ANALYSIS

We ended *Understanding the FFT* with a discussion of convolution and filters—and that leads us directly to the topic of network analysis, filters and, eventually, transient analysis. We assume anyone who's interested in this will be familiar with circuit analysis, but we'll review the fundamentals just to get everyone started on the same foot.[1]

4.1 NETWORK ANALYSIS

1. The voltage drop across a resistor is given by the relationship: $E = IR$. This, of course, is Ohm's law—the most fundamental equation of electrical circuits.

2. The sum of the voltage drops across the "load" side of a circuit must equal the applied voltage. When a voltage is applied to a closed circuit it causes a current to flow and, by 1. above, a current through a resistance "drops" a voltage. The applied voltage *forces* a current through the circuit and the *dropped* voltage *pushes back* against the applied force; therefore, the current *must* become steady when the voltage dropped equals the voltage applied (i.e., reaction equals action).

3. Several current carrying *legs* may tie together at a *node* (Fig. 4.1), but *there can be only one voltage at any node*! Similarly, there may be several circuit elements in any leg (i.e., simple series circuit), and consequently several voltages associated with these elements, but *there can be only one current flowing in any leg*.

With the above fundamentals we can write equations to determine the current that will flow, and the voltage drops that will result, in terms of the voltage applied to a circuit and the voltage dropping elements of the circuit.

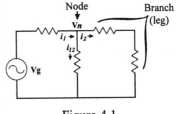

Figure 4.1

---

[1] Electrical engineers will be old friends with the frequency domain. Much of their education is presented directly in the frequency domain, and so they are pre-disposed to accept the reality of Fourier transforms.

4.2 THE VOLTAGE DIVIDER

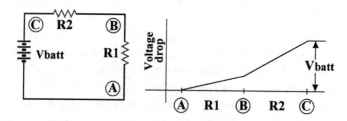

Figure 4.2 - Voltage Drop

We may write the following equation for the circuit of Figure 4.2:

$$V_{BB} = V_{R1} + V_{R2} \quad (V_{BB} = Vbatt) \quad\text{------} \quad (4.2A)$$
$$= I_1 R_1 + I_1 R_2 \quad\text{----------------} \quad (4.2B)$$
$$= I_1(R_1 + R_2) \quad\text{----------------} \quad (4.2C)$$

That is, the sum of the voltage drops across the resistors equals the applied voltage (i.e., the battery voltage) or, stated the other way around, when the battery voltage is included on the right hand of the equal sign, the voltage drops around the loop sum to zero:

$$0 = V_{R1} + V_{R2} - V_{BB} \quad\text{-------------} \quad (4.2D)$$

Now, most often, we would know the voltage applied and the values of the resistors, and would determine the current by an *inversion* of Eqn. (4.2C):

$$I_1 = V_{BB}/(R_1 + R_2) \quad\text{---------------} \quad (4.3)$$

We could then determine the voltage drop across either of the two resistors by simply applying Ohm's law. The configuration of Fig. 4.2 is commonly used to attenuate a signal and is referred to as a "voltage divider" (i.e., the two resistors *divide* the applied voltage between themselves). The voltage across R_1 and R_2 is given by:

$$V_{R1} = I_1 R_1 = [V_{BB}/(R_1 + R_2)] R_1 \quad\text{---------} \quad (4.4)$$
$$V_{R2} = I_1 R_2 = [V_{BB}/(R_1 + R_2)] R_2 \quad\text{---------} \quad (4.4A)$$

so that:

$$V_{R1}/V_{R2} = I_1 R_1 / I_1 R_2 = R_1/R_2 \quad\text{-----------} \quad (4.4B)$$

Network Analysis

4.3 MESH ANALYSIS

While many circuits take on a "half ladder" appearance (left, Fig. 4.3), configurations such as shown on the right side of Fig. 4.3 are not uncommon. *Network analysis* of complicated circuits such as these (i.e., the determination of currents and voltages in the circuits) is accomplished

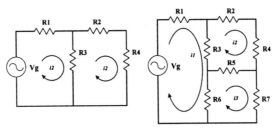

Figure 4.3 - Circuit Configurations

via simultaneous equations. The equations themselves are based on the fundamental principles stated above plus Kirchoff's laws.[2] There are two laws—one says the sum of currents entering and leaving any *node* must be zero and the other says the sum of the voltage drops around any loop must be zero. These observations are, of course, very fundamental and apply to both DC and AC circuits. [If we start from any node, and sum all the voltage *drops* around any loop, it's apparent we will end up at the same node and the same voltage. The sum of the voltage drops, no matter what path we take, *must therefore be equal to zero*. Similarly, since the current entering a node must equal the currents departing that node, the sum of the currents at any node must be zero. These bits of insight are Kirchoff's voltage and current laws (published in 1848).]

In Fig. 4.4 there are three loops (or meshes). To write our equations we *idealize* the current flowing in each loop as an independent current. Under this condition the *real* current flowing through R_{12} is equal to the *difference* between I_1 and I_2 (since *both independent currents* flow through this resistor). R_{12} is therefore called the *mutual impedance* between loops 1 and 2. A similar situation exists for the current flowing through R_{23} between loops 2 and 3. The *mesh impedance* of any loop is the sum of all the impedances through which the mesh current flows (e.g., the

---

[2]Gustav Robert Kirchoff (pronounced *Kirk' hof*) another "world-class" German physicist (1824-1887).

Mesh 1 impedance is equal to $R_1 + R_{12}$). The set of simultaneous equations is generated by summing the voltage drop around each mesh (i.e., the applied voltage[3] equals the voltage drops due to the mesh current flowing through the mesh impedance, minus the voltage drops introduced via the mutual impedances and currents in the other meshes; therefore:

$$V_g = I_1(R_1 + R_{12}) \quad -I_2(R_{12}) \quad -I_3(0) \quad (4.5A)$$
$$0 = -I_1(R_{12}) \quad +I_2(R_{12}+R_2+R_{23}) \quad -I_3(R_{23}) \quad (4.5B)$$
$$0 = -I_1(0) \quad -I_2(R_{23}) \quad +I_3(R_{23}+R_3+R_4) \quad (4.5C)$$

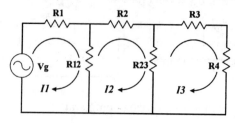

Figure 4.4 - Mesh Analysis

This yields three simultaneous independent equations in three unknowns which may always be solved for the unknown currents. A great deal of computation is involved and today circuit analysis is done almost exclusively on computers. We will look at this shortly, but first we must talk about "AC" circuits and the unique characteristics peculiar to this aspect of circuit analysis.[4]

4.4 AC CIRCUIT ANALYSIS

The "AC" stands for *Alternating Current* and generally refers to a *sinusoidal* driving signal (i.e., sinusoidal voltage source) and a sinusoidal response (i.e., a sinusoidal current). Besides electrical *resistance*, there are two other fundamental *elements* in ac electrical circuits that limit current—inductance and capacitance. These elements are characterized as follows: while the voltage across a resistor is proportional to the current flowing through it, the voltage across an inductor is proportional to the *rate*

---

[3] A voltage source is always indicated as a simple voltage generator and an internal impedance. It makes no difference whether the voltage is generated by thermal activity (noise), chemical activity (battery), or whatever—voltage is voltage.

[4] What we have been considering so far is DC (i.e., direct current) analysis which is a steady currents that flows when we use a battery or DC power supply.

Network Analysis 35

of change (i.e., the derivative) of the current flowing through it, and the voltage across a capacitor is proportional to the *accumulation* (i.e., the integral) of the current flowing into it (or, if you prefer, the *charge*):

$$V_L = L\, di/dt \quad \text{and} \quad V_C = 1/C \int i\, dt \quad (4.6A\&B)$$

where: L = value of inductance (in Henrys[5])
V_L = voltage dropped across an inductor
C = value of capacitance (in Farads[5A])
V_C = voltage dropped across a capacitor

It's remarkable that these three elements are *the* fundamental elements of passive, linear, time-invariant circuits.[6] The above equations are for the *time domain* response of these elements, but undergraduate students are taught about these components in the frequency domain.

4.5 OHM'S LAW FOR AC CIRCUITS

Ohm's law defines *resistance* as the ratio of voltage to current. For an ac current, at any given frequency, inductors and capacitors yield a constant ratio of ac voltage to ac current[7]—so they will obey this same relationship. Still, the behavior of these phase shifting, frequency dependent components is quite a bit different than resistors; and so, instead of calling their opposition to current flow *resistance*, we use the term *reactance*. The reactance equations are:

$$X_L = j2\pi f L \quad (4.7)$$
and: $$X_C = 1/j2\pi f C \quad (4.8)$$

where: X_L = inductive reactance
X_C = capacitive reactance
L = inductance
C = capacitance
$j = i = \sqrt{-1}$

---

[5]For Joseph Henry. Born in 1797, fourteen years after the American Revolution, he was one of the first world class American scientists (after Franklin).
[5A]Named in honor of the illustrious Michael Faraday, of course.
[6]See Chapter 2, Section 2.2.
[7]For sinusoids at any given frequency, it's obvious the rate of change is proportional to amplitude. The quantity of charge accumulated over any half cycle is likewise obviously proportional to the amplitude and, being *discharged* on the second half cycle, must necessarily yield an *ac current* proportional to amplitude.

The j accounts for the 90° phase shift, of course, and we combine these reactive elements with resistance just as we combine real numbers with imaginary numbers. The resulting complex quantity is given the name *impedance* and is designated by the symbol Z. For ac circuits, impedance

$$Z = R + j(2\pi fL - 1/(2\pi fC)) \qquad (4.9)$$

where: $1/j = 1/\sqrt{-1} = (-1 \cdot \sqrt{-1} \cdot \sqrt{-1})/\sqrt{-1} = -1 \cdot \sqrt{-1} = -j$

is used just as resistance is for dc circuits. We may plug these expressions for impedance directly into the simultaneous equations discussed earlier and analyze AC networks (but, of course, we must use complex arithmetic and solve the equations on a frequency by frequency basis).

But how, exactly, do we come by Eqns. (4.7) to (4.9)? First of all, *these equations refer to the behavior of circuits when driven with sinusoidal voltages.* We obtain Eqns. (4.7) and (4.9) by substituting sinusoidal driving functions into Eqns. (4.6 A&B).

$$I_L = I \sin(\omega t) \qquad (4.10)$$

where: I_L = current through the inductor
I = peak value of the sinusoid
$\omega = 2\pi f = 2\pi/T_0$ = angular velocity

then:
$$V_L = L \, di/dt = L \, d(I \sin(\omega t))/dt = \omega L \cdot I \cos(\omega t)$$
$$= j\omega L \cdot I \sin(\omega t)$$

and:
$$X_L = V_L/I_L = V_L/I \sin(\omega t) = j\omega L = j2\pi fL \qquad (4.7A)$$

We may derive Eqn. (4.8) in exactly the same way. The important thing to note here is that we have now moved our analysis onto a sinusoidal basis function—we are analyzing what happens when we drive our circuits with sinusoids. As we know, sinusoids represent a special case, and a sinusoid applied to the input of a linear circuit will always yield a sinusoidal response. It's apparent, then, the selection of sinusoids as the basis function for network analysis is not arbitrary. More intriguing still is the similarity of sinusoids as the basis function in Fourier analysis—in the next two chapters we shall expand on this similarity.

CHAPTER 5

AC CIRCUIT ANALYSIS PROGRAM

    We will need a program to solve the simultaneous equations discussed in Chapter 4. Again, you may have no interest in a program to do network analysis and, so long as you understand we will use this program to obtain the steady state transfer function of circuits in the next chapter, you may skip to the next chapter.

    The overall operation of the program is as follows: At line 10 we begin initializing, generating constants, and getting user defined constants such as frequency range, frequency steps and number of meshes. [Note: the number of meshes (which we call KB) determines the number of equations and the number of terms in each equation]. At line 120 we jump down to the circuit input routine where all the self and mutual impedances are input. At line 130 we jump down to calculate the transfer function of the circuit—the business end of the program. At line 140 we print the *Output Menu* which allows either drawing the transfer function on the computer screen or printing out the attenuation/phase vs. frequency to a printer...nothin' to this job.

```
'   ************************************************
'   *       ACNET 05-01 - NETWORK MESH ANALYSIS           *
'   ************************************************
10 SCREEN 9, 1: COLOR 15, 1: CLS
12 LOCATE 1, 20: PRINT "NETWORK MESH ANALYSIS": PRINT : PRINT
20 PI = 3.141592653589793#: PI2 = 2 * PI: ERFLG = 0
30 PRINT "A UNIT VOLTAGE SOURCE IS ASSUMED FOR MESH 1"
40 INPUT "ENTER DESIRED FREQUENCY RANGE (FMIN, FMAX)"; FMN, FMX
50 INPUT "ENTER NUMBER OF FREQUENCY STEPS PER DECADE"; DELF
60 FBOT = INT(LOG(FMN) / LOG(10))  ' COMMON LOG FMN
70 FTOP = INT(LOG(FMX) / LOG(10))  ' COMMON LOG FMX
80 IF LOG(FMX) / LOG(10) - FTOP > .000001 THEN FTOP = FTOP + 1
90 FSTEP = 10 ^ (1 / DELF)  ' FREQ STEP MULTIPLIER
100 INPUT "ENTER THE NUMBER OF MESHES"; KB
110 DIM COMP(3, KB, KB), Z(2, KB, KB+1), F(2,((FTOP-FBOT)*DELF)+1)
120 GOSUB 20000  ' INPUT CIRCUIT ELEMENTS
130 GOSUB 10000  ' CALCULATE XFER IMPEDANCE
140 CLS : LOCATE 2, 20: PRINT "OUTPUT MENU"
150 PRINT : PRINT "  1 = PLOT DATA"
155 PRINT : PRINT "  2 = PRINT DATA (TO SCREEN)"
160 PRINT : PRINT "  3 = OUTPUT TO PRINTER"
170 LOCATE 18, 3: PRINT "9 = EXIT"
180 PRINT : PRINT "     INPUT SELECTON"
190 A$ = INKEY$: IF A$ = "" THEN 190
200 ON VAL(A$) GOSUB 800, 10090, 600, 235, 235, 235, 235, 235, 240
```

```
230 GOTO 140
235 RETURN
240 END

' **************************************
' *       PRINT OUTPUT TO PRINTER       *
' **************************************
600 INPUT "CIRCUIT NAME "; A$
605 LPRINT DATE$, TIME$: LPRINT : LPRINT A$: LPRINT : LPRINT
610 F0 = 10 ^ FBOT: F = F0: PLC = 52
615 FOR I = 1 TO ((FTOP - FBOT) * DELF) + 1
620 LPRINT USING "#####.##_         "; F; ' PRINT FREQUENCY
625 LPRINT USING "##.####_ "; F(1, I) * RTERM; ' PRINT VOLTAGE
630 J$ = "+j ": IF SGN(F(2, I)) < 0 THEN J$ = "-j "
635 LPRINT J$;: LPRINT USING "##.####_      "; ABS(F(2, I)*RTERM);
640 MAG = SQR(F(1, I) ^ 2 + F(2, I) ^ 2)
642 IF MAG = 0 THEN LPRINT "0.00000 / 0.00": GOTO 660
644 LPRINT USING "##.#####_ "; RTERM * MAG; : LPRINT " /";
646 IF F(1, I) = 0 AND F(2, I) >= 0 THEN BTA = PI / 2: GOTO 655
648 IF F(1, I) = 0 AND F(2, I) < 0 THEN BTA = -PI / 2: GOTO 655
650 BTA = ATN(F(2, I) / F(1, I)): IF F(1, I)<0 THEN BTA = BTA+PI
655 LPRINT USING "###.##"; BTA * 180 / PI ' PRINT DEGREES PHASE
660 IF I < PLC THEN 690 ' CHK END OF PAGE
665 LPRINT "                          (MORE)"
670 LPRINT CHR$(12) ' FORM FEED
675 PLC = PLC + 52 ' SET END OF NEXT PAGE
680 FOR LCTR = 1 TO 3: LPRINT : NEXT ' PRINT HEAD SPACE
690 F = F0 * FSTEP ^ I' SET NEXT FREQUENCY
692 NEXT I ' PRINT NEXT LINE
694 LPRINT CHR$(12) ' FORM FEED FINAL PAGE
698 RETURN

' **************************************
' *          PLOT XFER FUNCTION         *
' **************************************
800 SCREEN 9, 1: COLOR 15, 1: CLS ' SELECT 640 x 350 RESOLUTION
810 X0 = 20: Y0 = 20: YX = 320: XX = 620 ' 600 x 300 GRAPH
812 LINE (X0, Y0)-(X0, YX) ' DRAW COORDINATES
814 LINE (X0, YX)-(XX, YX)
820 FREQNO = (FTOP - FBOT) * DELF: YMAX = 0
822 FOR I = 1 TO FREQNO
824 Y1 = SQR(F(1, I) ^ 2 + F(2, I) ^ 2)
826 IF Y1 > YMAX THEN YMAX = Y1
828 NEXT I
830 KX = 600 / FREQNO: KY = 200 / (YMAX * RTERM)
832 LOCATE 8, 1: PRINT YMAX * RTERM
834 Y1 = RTERM * SQR(F(1, 1) ^ 2 + F(2, 1) ^ 2)
836 LINE (X0, YX - (KY * Y1))-(X0, YX - (KY * Y1))
840 FOR I = 1 TO FREQNO
842 Y1 = RTERM * SQR(F(1, I) ^ 2 + F(2, I) ^ 2)
```

Circuit Analysis Program

```
844 LINE -(X0 + (I - 1) * KX, YX - (KY * Y1))
846 NEXT
848 LOCATE 24, 2: PRINT 10^FBOT; : LOCATE 24,73: PRINT 10^FTOP;
850 LOCATE 25, 30: PRINT "FREQUENCY"; : LOCATE 25, 2
852 INPUT A$
854 RETURN
' ***********************************
' *       CALCULATE XFER FUNCTION       *
' ***********************************
10000 F0 = 10 ^ FBOT: F = F0: FREQ = 0 ' FREQ = DATA POINT COUNTER
10010 GOSUB 11000 ' CALC IMPEDANCE MATRIX
10020 GOSUB 12000 ' SOLVE FOR CURRENTS
10040 ' GOSUB 13000 ' PRINT OUT CURRENTS
10050 FREQ = FREQ + 1 ' SET FOR NEXT DATA POINT
10060 F(1, FREQ) = Z(1, KB, KB + 1): F(2, FREQ) = Z(2,KB,KB+1) ' SAVE
10070 F = F0 * FSTEP ^ FREQ ' GET NEXT FREQUENCY
10080 IF F <= FMX + .1 THEN 10010 ' JOB DONE?
10090 CLS : PLC = 20 ' CLEAR SCREEN FOR PRINTOUT
10100 F = 10 ^ FBOT ' GET BOTTOM FREQUENCY
10110 FOR I = 1 TO ((FTOP - FBOT) * DELF) + 1
10120 PRINT USING "#####.##_   "; F; ' PRINT FREQUENCY
10130 PRINT USING "##.####_   "; F(1, I) * RTERM; ' ? REAL COMPONENT
10140 J$ = "+j ": IF SGN(F(2, I)) < 0 THEN J$ = "-j "
10150 PRINT " "; J$;
10160 PRINT USING "#.####_   "; ABS(F(2,I)*RTERM); 'PRINT IMAGINARY
10170 MAG = SQR(F(1, I) ^ 2 + F(2, I) ^ 2) ' FIND MAGNITUDE
10180 IF MAG = 0 THEN PRINT "0.0000 /0.00": GOTO 10240
10190 PRINT USING "#.##### "; RTERM * MAG; : PRINT " /";
10200 IF F(1, I) = 0 AND F(2, I) >= 0 THEN BTA = PI / 2: GOTO 10230
10210 IF F(1, I) = 0 AND F(2, I) < 0 THEN BTA = -PI / 2: GOTO 10230
10220 BTA = ATN(F(2, I)/ F(1, I)): IF F(1, I)<0 THEN BTA = BTA+PI
10230 PRINT USING "###.##"; BTA * 180 / PI ' PRINT PHASE IN DEG.
10240 IF I < PLC THEN 10270 ' SCREEN FULL?
10250 PLC = PLC + 20: PRINT "                        (MORE)"
10260 INPUT "ENTER TO CONTINUE"; A$ ' WAIT FOR USER
10270 F = F0 * FSTEP ^ I' NEXT FREQ
10280 NEXT I ' PRINT NEXT COMPONENT
10290 INPUT "ENTER TO CONTINUE"; A$ ' WAIT
10300 RETURN ' WERE OUTA HERE

' ***********************************
' *       COMPUTE IMPEDANCE       *
' ***********************************
11000 CLS : WF = PI2 * F ' GET ANGULAR FREQUENCY
11010 FOR I = 1 TO KB ' FOR ALL MESHES
11020 Z(1, I, I) = COMP(1, I, I) ' GET RESISTOR VALUE
11030 IF COMP(3,I,I) = 0 THEN Z(2,I,I) = WF*COMP(2,I,I): GOTO 11060
11040 IF F = 0 THEN Z(2,I,I) = -1E+18:GOTO 11060
11050 Z(2, I, I) = WF * COMP(2, I, I) - 1 / (WF * COMP(3, I, I))
11060 FOR J = I + 1 TO KB ' GET MUTUAL IMPEDANCES
```

```
11070 Z(1, I, J) = -COMP(1, I, J) ' MUTUAL RESISTANCE
11080 IF COMP(3,I,J) = 0 THEN Z(2,I,J) = -WF*COMP(2,I,J): GOTO 11110
11090 IF F = 0 THEN Z(2,I,J) = 1E+18: GOTO 11110 ' FUDGE DC CAP
11100 Z(2, I, J) = -(WF * COMP(2, I, J) - 1 / (WF * COMP(3, I, J)))
11110 Z(2, J, I) = Z(2, I, J): Z(1, J, I) = Z(1, I, J)
11120 NEXT J ' NEXT MUTUAL IMPEDANCE
11130 NEXT I ' NEXT MESH
11140 Z(1, 1, KB + 1) = 1: Z(2, 1, KB + 1) = 0 ' DRIVE POINT VOLTAGE
11150 FOR I = 2 TO KB ' SET ALL OTHER MESH VOLTAGE SOURCES (ZERO)
11160 Z(1, I, KB + 1) = 0: Z(2, I, KB + 1) = 0 ' ALL ZERO
11170 NEXT I
11250 RETURN

' ************************************
' *           SOLVE FOR CURRENTS     *
' ************************************
12000 FOR J = 1 TO KB ' STEP THROUGH ALL DIAGONAL LOCATIONS
12200 ' GET A "1" COEFFICIENT ON THE DIAGONAL
12210 IF Z(2, J, J) = 0 THEN ZRE = 1 / Z(1, J, J): ZIM = 0: GOTO 12300
12220 IF Z(1,J,J) = 0 THEN ZRE = 0: ZIM = -1 / Z(2,J,J): GOTO 12300
12230 BTA = ATN(Z(2,J,J)/Z(1,J,J)):IF Z(1,J,J)<0 THEN BTA = BTA+PI
12240 BTA = .5 * SIN(2 * BTA)
12250 ZRE = BTA / Z(2, J, J): ZIM = -BTA /Z(1,J,J)' INVERT Z(x,J,J)
12300 FOR K = J TO KB + 1
12310 Z1 = Z(1,J,K) * ZRE - (Z(2, J, K) * ZIM)
12315 Z(2,J,K) = Z(1,J,K) * ZIM + (Z(2,J,K) * ZRE): Z(1,J,K) = Z1
12320 NEXT K
12330 FOR I = 1 TO KB ' GET 0 IN THIS COL. FOR ALL OTHER EQNS.
12340 IF I = J THEN 12400
12350 XT = Z(1, I, J): YT = Z(2, I, J): IF XT =0 AND YT =0 THEN 12400
12360 FOR K = J + 1 TO KB + 1' DON'T BOTHER WITH "ZERO" POSITIONS
12370 Z(1, I, K) = Z(1, I, K) - (Z(1, J, K) * XT - (Z(2, J, K) * YT))
12380 Z(2, I, K) = Z(2, I, K) - (Z(1, J, K) * YT + (Z(2, J, K) * XT))
12390 NEXT K
12400 NEXT I
12410 NEXT J ' REPEAT FOR ALL DIAGONAL POSITIONS.
12420 RETURN

' ************************************
' *          PRINTOUT CURRENTS       *
' ************************************
13000 CLS : PRINT
13010 FOR I = 1 TO KB
13020 PRINT "I"; I; "=";
13030 PRINT USING "###.###"; Z(1, I, KB + 1);
13040 PRINT "+j";
13050 PRINT USING "###.###"; Z(2, I, KB + 1)
13060 NEXT I
13080 RETURN
```

Circuit Analysis Program

```
'   ************************************
'   *        INPUT CIRCUIT ELEMENTS        *
'   ************************************
20000 PRINT : PRINT "       INPUT CIRCUIT ELEMENTS"
20010 PRINT : GOSUB 21000 ' PRINT INSTRUCTIONS
20020 FOR I = 1 TO KB
20030 PRINT : PRINT "INPUT VALUES FOR MESH "; I: PRINT
20040 PRINT "ENTER MESH SELF RESISTANCE, INDUCTANCE, CAPACITANCE";
20050 INPUT COMP(1, I, I), COMP(2, I, I), COMP(3, I, I)
20060 FOR J = I + 1 TO KB
20070 PRINT "ENTER MESH ";
20080 PRINT USING "##_ "; I; J;
20090 PRINT " MUTUAL RESISTANCE, INDUCTANCE, CAPACITANCE";
20100 INPUT COMP(1, I, J), COMP(2, I, J), COMP(3, I, J)
20110 COMP(1,J,I)=COMP(1,I,J):COMP(2,J,I)=COMP(2,I,J)
20125 COMP(3, J, I) = COMP(3, I, J)
20130 NEXT J
20140 NEXT I
20150 INPUT "ENTER TERMINATION RESISTANCE"; RTERM
20160 RETURN
'   ****************************************
'   *               INSTRUCTIONS               *
'   ****************************************
21000 PRINT "1) LAYOUT CIRCUIT INTO MESHES"
21010 PRINT "2) MESH SELF IMPEDANCE = ";
21015 PRINT "SERIES SUMMATION OF ALL IMPEDANCES IN LOOP"
21020 PRINT "3) MESH MUTUAL IMPEDANCE = ";
21025 PRINT "COMMON IMPEDANCES BETWEEN LOOPS": PRINT
21030 INPUT "ENTER TO CONTINUE"; A$
21040 RETURN
```

5.1 CIRCUIT INPUT

A note about data input is in order. All mesh self-impedances and mutual impedances must be reduced to a single *series string* of resistance, inductance and capacitance. When summing inductors and resistors within a loop, simply add them, but series capacitors add as:

$$C_T = C_1 C_2 / (C_1 + C_1) = 1/[(1/C_1) + (1/C_2) + ...]$$

A 15 mh (milli-henry) inductor is input as 0.015 and a 12.25μfd (microfarad) capacitor is input as 12.25E-6. In all cases the mesh and mutual impedances are input as R,L,C (e.g., we input 1000 ohms resistance, 15 mh inductance, and 12.25 μfd capacitance as 1000, 0.015, 12.25E-6).

For example, we would input the following six pole Butterworth circuit as follows: mesh 1 self impedance (Z11) is input as 0, 15.47, 17.42E-6 (i.e., zero resistance, 15.47 hy of inductance, and 17.42 μfd of

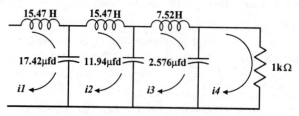

Figure 5.1 - 6 Pole Butterworth (fco = 16 Hz)

capacitance). The mutual impedance between meshes 1 and 2 (Z12) is input 0, 0, 17.42E-6. Z13 is, of course, 0, 0, 0 because there is no mutual impedance between these loops (that is, the current flowing in loop 3 will introduce no voltage into loop 1, so we multiply I_3 by zero in the loop 1 equation. Similarly with Z14. We now come to the self impedance of loop 2 (Z22) and here we have 0, 15.47, and <u>7.082E-6</u>. The 7.082E-6 is the *series sum of the two capacitors in this loop!* Z23 is 0, 0, 11.94E-6 and Z24 is, again, 0, 0, 0. Okay, you know how to input the rest of the data. [Note that we didn't have to input Z21 because it's identical to Z12. You also recognize that the *terminating resistance* is 1000 ohms.]

No provision is made for entering voltage *sources* into any of the loops (since our primary intention is to analyze filter circuits), but this provision can easily be added (lines 11150 - 11170). We also assume that the output will be taken from the last loop (not necessarily true in the general case). Furthermore, to get the output in terms of voltage, we must input the *termination resistance* mentioned above. That is, this program solves for mesh currents and, to get things back in terms of voltage, the output mesh current must be multiplied by the filter terminating resistance (see for example, lines 625, 635, 644, 832, 834, 842, 10130, 10160 or 10190). If you want to see the *current*, simply make this value equal to 1.0. The frequency range must be specified (in Hz—line 40) and the number of steps within each decade of frequency must also be specified (typically 8 to 100—line 50). In this program the frequencies are computed logarithmically to keep the data from becoming excessive; however, when using this program with the FFT we will have to calculate frequency in an arithmetic progression.

Circuit Analysis Program

5.2 SOLVING THE SIMULTANEOUS EQUATIONS

The heart of this program is the subroutine located at 12000 ***SOLVE FOR CURRENTS***. [Note: in electrical engineering the standard technique for solving simultaneous equations is *Determinants*, which is a side-street off of the main avenue of *Matrix Algebra*. We will *not* use determinants here but tackle the simultaneous equations directly.]

Our objective is to find the loop currents of Equations (4.5). We will do this by reducing the coefficients of all but the *self-impedance loop current* to zero in each equation (i.e., we reduce the equations to the form:

$$Vx/Zx = I_1 \quad +0 \quad +0 \quad (5.4A)$$
$$Vy/Zy = 0 \quad +I_2 \quad +0 \quad (5.4B)$$
$$Vz/Vz = 0 \quad +0 \quad +I_3 \quad (5.4C)$$

where Vx, Zx, Vy, Zy, etc., will be determined in the process. To do this we employ only the rules that we may *multiply* both sides of an equation by equals without changing the equation, and we may *add* equals to both sides of an equation without changing the equality. We start with:

$$V1 = I_1(Z11) \quad -I_2(Z12) \quad -I_3(Z13) \quad (5.5A)$$
$$V2 = -I_1(Z21) \quad +I_2(Z22) \quad -I_3(Z23) \quad (5.5B)$$
$$V3 = -I_1(Z31) \quad -I_2(Z32) \quad +I_3(Z33) \quad (5.5C)$$

where Z11, Z22 and Z33 (along the diagonal) are the *self impedances* of the loops and Z12, Z13, etc., are the *mutual impedances* between the loops as explained earlier. Very well then, we start by dividing both sides of Eqn. (5.5A) by Z11. V1 becomes V1/Z11, the coefficient of I_1 becomes unity, the coefficient of I_2 becomes Z12/Z11 and the coefficient of I_3 becomes Z13/Z11:

$$V1/Z11 = I_1 \quad -I_2(Z12/Z11) \quad -I_3(Z13/Z11) \quad (5.5A')$$

Next we turn to Eqn. (5.5B). Here we want to reduce the Z21 coefficient to zero, and we can do this by multiplying a copy of Eqn. (5.5A') by Z21 and subtracting the result from (5.5B). Since the coefficient of I_1 has been normalized to 1.0, Eqn. (5.5B) will become:

$$V2\text{-}(V1'*Z21) = 0 \quad +I_2(Z22\text{-}(Z12'*Z21)) \quad -I_3(Z23\text{-}(Z13'*Z21)) \quad (5.5B')$$

where $V1'$, $Z12'$ and $Z13'$ are the coefficients shown in Eqn. (5.5A'). We then do the same thing to Eqn.(5.5C) (i.e., we reduce Z31 to zero by subtracting a copy of (5.5A) multiplied by Z31) yielding:

$$V1' = I_1 \quad -I_2(Z12') \quad -I_3(Z13') \tag{5.6A}$$
$$V2' = 0 \quad +I_2(Z22') \quad -I_3(Z23') \tag{5.6B}$$
$$V3' = 0 \quad -I_2(Z32') \quad +I_3(Z33') \tag{5.6C}$$

We now turn to Equation (5.6B) and divide all of its terms (on both sides) by $Z22'$, normalizing the I_2 coefficient to unity:

$$V2'' = 0 \quad +I_2 \quad -I_3(Z23'/Z22') \tag{5.5B''}$$

We then subtract replicas of this equation (multiplied by $Z12'$ and $Z32'$ respectively) from Eqns.(5.6A) and (5.6C) resulting in:

$$V1'' = I_1 \quad 0 \quad -I_3(Z13'') \tag{5.7A}$$
$$V2'' = 0 \quad +I_2 \quad -I_3(Z23'') \tag{5.7B}$$
$$V3'' = 0 \quad 0 \quad +I_3(Z33'') \tag{5.7C}$$

Finally, we normalize $Z33''$, subtract from the other two eqns., and:

$$V1''' = I_1 \quad 0 \quad 0 \tag{5.8A}$$
$$V2''' = 0 \quad +I_2 \quad 0 \tag{5.8B}$$
$$V3''' = 0 \quad 0 \quad +I_3 \tag{5.8C}$$

This sort of manipulation is suggestive of Determinant mechanics *(studying Determinants is at least as much fun as participating in a peanut butter eating race)*. One advantage to *this* algorithm is that it simultaneously obtains a solution for *all* of the unknowns (which may occasionally be useful). It can be speeded up by solving only for the output mesh current, but the speed is only doubled (approximately).

Let's look at this exceedingly simple-minded program. We will step (line 12000) through the diagonal elements $Z(x,J,J)$ [*x designates the real/imaginary side of the impedance*] of the equations with J specifying row *and* column. First we normalize this coefficient to a value of 1.0—that is, we obtain the reciprocal of the diagonal component (lines 12210 - 12250) and multiply this through the equation (lines 12300 - 12320). In our case, it will frequently occur that either the real or

Circuit Analysis Program

imaginary component will be zero and, if this is so, our task is greatly simplified (lines 12210 and 12220)—otherwise, there is nothing left but to crunch the numbers. Note that complex quantities make everything complicated here. To obtain the reciprocal of a complex number we must convert to polar coordinates; however, since we need the result back to rectangular format, we take short-cuts (lines 12230 - 12250).

```
12000 FOR J = 1 TO KB ' STEP THROUGH DIAGONAL - KB = # OF EQNS.
12200 'GET A "1" COEFFICIENT ON THE DIAGONAL
12202 'Z(2,J,J) = IMAGINARY PART OF IMPEDANCE & Z(1,J,J) = REAL PART
12210 IF Z(2, J, J) = 0 THEN ZRE = 1 / Z(1, J, J): ZIM = 0: GOTO 12300
12220 IF Z(1,J,J) = 0 THEN ZRE = 0: ZIM = -1 / Z(2,J,J): GOTO 12300
12230 BTA = ATN(Z(2,J,J)/Z(1,J,J)):IF Z(1,J,J)<0 THEN BTA = BTA+PI
12240 BTA = .5 * SIN(2 * BTA)
12250 ZRE = BTA / Z(2, J, J): ZIM = -BTA /Z(1,J,J)' INVERT Z(x,J,J)
12300 FOR K = J TO KB + 1
12310 Z1 = Z(1,J,K) * ZRE - (Z(2, J, K) * ZIM)
12315 Z(2,J,K) = Z(1,J,K) * ZIM + (Z(2,J,K) * ZRE): Z(1,J,K) = Z1
12320 NEXT K
```

Once we have a "1" on the diagonal we must then get a "0" in this column for all other equations. We do this, as we said, by subtracting a copy of the normalized equation (after multiplying by the coefficient in the Jth column) from each equation (lines 12330 - 12410). We do not want to subtract the normalized equation from *itself*, however, as that would get a zero on the diagonal; so, when I = J (line 12340) we skip this routine. Furthermore, we can speed things up considerably by avoiding needless computation so, if the element is already zero, we forego all the arithmetic (end of line 12350). Also, we know that all the elements preceding this column are already zero, and all the elements in this column will be zero after we perform this routine, so we needn't actually do the arithmetic on any columns to the left of J + 1 (line 12360). [Note: all voltage sources are located at row position KB + 1—see routine at lines 11140-11170]

```
12330 FOR I = 1 TO KB ' GET 0 IN THIS COL. FOR ALL OTHER EQNS.
12340 IF I = J THEN 12400
12350 XT = Z(1, I, J): YT = Z(2, I, J): IF XT = 0 AND YT = 0 THEN 12400
12360 FOR K = J + 1 TO KB + 1' DON'T BOTHER WITH "ZERO" POSITIONS
12370 Z(1, I, K) = Z(1, I, K) - (Z(1, J, K) * XT - (Z(2, J, K) * YT))
12380 Z(2, I, K) = Z(2, I, K) - (Z(1, J, K) * YT + (Z(2, J, K) * XT))
12390 NEXT K
12400 NEXT I
12410 NEXT J ' REPEAT FOR ALL DIAGONAL POSITIONS.
12420 RETURN
```

When we come out of this routine we will have the mesh current (for every loop) in the KB+1 location of each row. [The *real* part of the current is located in the "1" side of the array and the *imaginary* part in the "2" side.]

The data output routines are pretty much what they appear to be so we will not take up time and pages with this trivia. You might want to run this program and analyze the Butterworth filter shown in Fig. 5.1—just to see if it works.

This program does what all network analysis programs do—it gives the transfer function of the circuit. If, some day, the big work station is in use, and you need a quick answer, it could come in handy. For us, however, its real purpose will become apparent in the next chapter.

CHAPTER 6

GIBBS, FILTERS, AND TRANSIENT ANALYSIS

The network analysis we have considered so far is called *Steady State Analysis*. It yields the response of a circuit to constant amplitude sinusoids. As often than not, however, we are interested in the *transient response* of the circuit—see Fig. 6.1. The term *transient response* refers to the response of a circuit to a sudden change in the input amplitude. In audio work, for example, an oscillatory response to a sudden change in amplitude can be very annoying—in servo systems transient response is paramount.

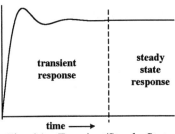

Fig. 6.1 - Transient/Steady State

6.1 BAND LIMITED SIGNALS

We know, via Fourier analysis, that we may decompose signals into their harmonic components. Now, the *size* of the frequency spectrum (i.e., the *width* of the *band of frequencies* that make up any given signal), is referred as its *bandwidth*. Theoretically, discontinuous waveforms contain an infinite number of components. A perfect square wave, for example, requires an infinite number of harmonics, and this implies an infinite bandwidth. When working with *real* square waves, however, we will always disregard the harmonics above some arbitrarily high value—after all, the harmonics are becoming smaller and smaller with increasing frequency. At some point their contribution to the total signal *must* become negligible.[1] By throwing away all frequency components above some *cutoff frequency* we *limit* the bandwidth and created a *band limited* signal. In practice, all physically real signals are band limited; but, in the art of engineering tradeoff and compromise, the limiting of bandwidth can be a ticklish job.

---

[1] This is true for all physically real wave forms. In fact, if you look closely at *any* real application you will almost certainly find no *infinity* is really involved. In practice, limits to resolution and dynamic range leave *infinities* "suppositions of our imaginations."

6.2 THE GIBBS[2] PHENOMENON (THEORY)

Our comment about throwing away the harmonics above some high frequency leads us to the Gibbs phenomenon. In its classical manifestation, Gibbs' phenomenon appears when we try to construct discontinuous waveforms by summing harmonics. For example, in constructing a square wave by summing the odd harmonics, we find that even after a good replica of a square wave has been obtained, there still exists a transient *overshoot* and *ringing* at the transitions (Fig. 6.2). If we add more harmonic components, the transient becomes shorter, but the amplitude remains approximately 8.95% of the transition. Let's use another computer program to illustrate this effect:

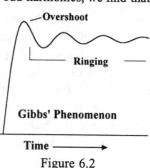

Figure 6.2

```
10 REM *** GENERATE SQUARE WAVE TRANSITION ***
12 INPUT "NUMBER OF TERMS";N'  SELECT NO. OF HARMONICS TO BE SUMMED
14 CLS
16 PRINT"TIME         AMPLITUDE"' PRINT DATA HEADER
18 PRINT
19 REM LINE 20 SETS PI, TIME INTERVAL, AND TIME STEP INCREMENT
20 PI = 3.14159265358#:K1=2*PI/N:K2=PI/(N*8)
30 FOR I = 0 TO K1 STEP K2 ' K1=TIME INTERVAL & K2=TIME INCREMENT
32 Y=0
40 FOR J=1 TO N STEP 2 ' N=NUMBER OF HARMONICS SUMMED
42 Y=Y+SIN(J*I)/J ' SUM IN HARMONIC CONTRIBUTION
44 NEXT J ' SUM ALL HARMONICS
50 PRINT USING "#.######_      ";I/PI,Y*4/PI:REM PRINT OUTPUT
60 NEXT I ' CALCULATE NEXT DATA POINT
70 INPUT A$ ' WAIT
```

Line 12 allows us to select the number of sine wave components to be summed into the square wave approximation. Lines 14 through 18 print the heading for our data printout. Line 20 defines the constants used in the program: K1 sets the distance we will investigate past the leading edge of the square wave. Note that this distance is made inversely proportional to the number of harmonic components summed—the reason

---

[2]Josiah Willard Gibbs (1839-1903) published a solution to this puzzle (*Nature*, 27 April, 1899) settling a dispute between A.A. Michelson and A.E.H. Love. Once again, however, the math had been worked out and published 50 years earlier by Henry Wilbraham.

Transient Analysis

will be apparent shortly. K2 selects the distance between successive data points on the time base. This particular selection of constants will always provide 16 data points which portray the leading edge of the square wave. Lines 30 through 60 perform the familiar routine of summing sinusoids to reconstruct a square wave. Line 30 sets up a loop which steps through the time increments I, and line 40 sets up a nested loop which sums in the value of each harmonic at that point.

FOR 100 COMPONENTS		FOR 1000 COMPONENTS	
TIME	AMPLITUDE	TIME	AMPLITUDE
0.000000	0.000000	0.000000	0.000000
0.001250	0.247868	0.000125	0.247868
0.002500	0.483181	0.000250	0.483179
0.003750	0.694527	0.000375	0.694522
0.005000	0.872665	0.000500	0.872654
0.006250	1.011305	0.000625	1.011288
0.007500	1.107583	0.000750	1.107558
0.008750	1.162170	0.000875	1.162140
0.010000	1.179013	0.001000	1.178981
0.011250	1.164758	0.001125	1.164728
0.012500	1.127916	0.001250	1.127892
0.013750	1.077879	0.001375	1.077871
0.015000	1.023911	0.001500	1.023921
0.016250	0.974208	0.001625	0.974237
0.017500	0.935151	0.001750	0.935197
0.018750	0.910810	0.001875	0.910869

If you run this program for 100 and 1000 frequency components, you will get the results shown above. Except for "scaling" of the time interval, there is little difference between these two runs. On adding more components still, we find the overshoot *remains* a constant 8.9% of the transition amplitude (the transition in the above example is from -1 to +1). *Only the time scale will change.*

This is known as the Gibbs phenomenon. The overshoot will apparently remain at 8.9%—even when the number of components increases without limit! This would seem to indicate a series of sinusoids will never converge to a perfect reproduction of a discontinuous function; still, we should not lose sight of the fact that as the number of components *approaches* infinity, even though the *amplitude* of the overshoot remains constant, the *duration* of the transient approaches zero, and zero duration implies non-existence.

It's sometimes imagined that the Gibbs phenomenon indicates a defect in the DFT/Inverse DFT process, but *you may transform and inverse transform any data (until the silicon in your chips turns back into sand), and you will always get the same waveform you started with!* This phenomenon concerns discontinuities (i.e., zero risetimes), and we will never encounter zero risetimes in physically realizable signals and systems. When considering *finite* risetimes we need only sample at rates that are high (with respect to the rise-time of the signal) and we will faithfully capture the finite number of harmonics in the signal (we will discuss sampling and sampling rate in Chapter 10).

6.3 THE PRACTICAL SIDE OF GIBBS

The reason why the Gibbs transient is so persistent in the above illustrations is that, in every case, the summation of sinusoids is *incomplete*. All of the components, from the harmonic at which we stop the summing process up to infinity, are missing. If we add more components we will smooth-out the present ripples, but since there is no end to this summing process, it must forever remain incomplete, and we will only find a new, higher frequency oscillation. Now, as idealists, we can hold every real square wave accountable for an infinite number of missing harmonic components (compared to an ideal, perfect, discontinuous wave). So, then, shouldn't we see Gibbs' phenomenon present in every real square wave? In fact, every time we filter-off the higher frequency components of a real square wave (surely duplicating the incomplete spectrum described above) the Gibbs phenomenon should become apparent at the transitions!

Does this actually happen? We can easily find out by taking the transform of a "perfect" square wave, truncating the upper harmonic components, and reconstructing the time domain function. We will need an FFT program to perform this exercise (and the remaining illustrations in this chapter); so, we will borrow the following program (taken intact from Appendix 10 of *Understanding the FFT*). We have added routines to generate our initial square wave function, of course, and to truncate the spectrum of the frequency domain function (as well as a couple of other methods for filtering the spectrum). Other than that, it's the program presented in Appendix 10 of *Understanding the FFT*.

Transient Analysis 51

6.4 THE FFT PROGRAM

```
10 '      *** (FFT06-01) FFT/INV FFT ***
11 ' THIS PROGRAM ILLUSTRATES GIBBS PHENOMENON IN PRACTICE
12 SCREEN 9, 1: COLOR 15, 1
13 CLS : PRINT : PRINT "INPUT NUMBER OF DATA POINTS AS 2^N"
14 INPUT "N = "; N
16 Q = 2 ^ N: Q1 = Q - 1: N1 = N - 1
18 Q2 = Q / 2: Q3 = Q2 - 1: Q4 = Q / 4: Q5 = Q4 - 1: Q8 = Q / 8
20 DIM C(Q), S(Q), KC(Q), KS(Q)
22 X0 = 50: Y0 = 200: XSF = 500 / Q: YSF = 1024: IOFLG = 1
23 PI = 3.141592653589793#: P2 = 2 * PI: K1 = P2 / Q: PI2 = P2

24 '   **** TWIDDLE FACTOR TABLE GENERATION ****
26 FOR I = 0 TO Q: KC(I) = COS(K1 * I): KS(I) = SIN(K1 * I)
28 IF ABS(KC(I)) < .0000005 THEN KC(I) = 0 ' CLEAN UP TABLES
30 IF ABS(KS(I)) < .0000005 THEN KS(I) = 0
32 NEXT I
34 FOR I = 1 TO Q1: INDX = 0
36 FOR J = 0 TO N1
38 IF I AND 2 ^ J THEN INDX = INDX + 2 ^ (N1 - J)
40 NEXT J
42 IF INDX > I THEN SWAP KC(I), KC(INDX): SWAP KS(I), KS(INDX)
44 NEXT I
45 '   ****  PRINT MAIN MENU  ****
46 CLS
50 PRINT SPC(30); "MAIN MENU": PRINT : PRINT
52 PRINT SPC(5); "1 = GENERATE STEP FUNCTION & TRANSFORM": PRINT
54 PRINT SPC(5); "2 = FORWARD TRANSFORM": PRINT
56 PRINT SPC(5); "3 = MODIFY SPECTRUM": PRINT
58 PRINT SPC(5); "4 = INVERSE TRANSFORM": PRINT
60 PRINT SPC(5); "5 = MODIFY SYSTEM": PRINT
62 PRINT SPC(5); "6 = EXIT": PRINT
70 PRINT SPC(10); "MAKE SELECTION :";
80 A$ = INKEY$: IF A$ = "" THEN 80
90 A = VAL(A$): ON A GOSUB 600, 160, 1000, 170, 700, 990
95 GOTO 46

100 '     ****  FFT ROUTINE  ****
106 T9 = TIMER
112 FOR M = 0 TO N1: QT = 2 ^ (N - M): QT1 = QT - 1
114 QT2 = QT / 2: QT3 = QT2 - 1: KT = 0
116 ' *** UNIVERSAL BUTTERFLY ***
118 FOR J = 0 TO Q1 STEP QT: KT2 = KT + 1
120 FOR I = 0 TO QT3: J1 = I + J: K = J1 + QT2
122 CTEMP = (C(J1) + C(K) * KC(KT) - K6 * S(K) * KS(KT)) / SK1
124 STEMP = (S(J1) + K6 * C(K) * KS(KT) + S(K) * KC(KT)) / SK1
126 CTEMP2 = (C(J1) + C(K) * KC(KT2) - K6 * S(K) * KS(KT2)) / SK1
128 S(K) = (S(J1) + K6 * C(K) * KS(KT2) + S(K) * KC(KT2)) / SK1
130 C(K) = CTEMP2: C(J1) = CTEMP: S(J1) = STEMP
```

```
132 NEXT I
134 KT = KT + 2: NEXT J
136 NEXT M
140 ' *** BIT REVERSAL FOR FINAL DATA ***
142 FOR I = 1 TO Q1: INDX = 0
144 FOR J = 0 TO N1
146 IF I AND 2 ^ J THEN INDX = INDX + 2 ^ (N1 - J)
148 NEXT J
150 IF INDX > I THEN SWAP C(I), C(INDX): SWAP S(I), S(INDX)
152 NEXT I
154 GOTO 200

158 ' *** FORWARD FFT ***
160 K6 = -1: SK1 = 2
162 CLS : HDR$ = "FREQ     F(COS)        F(SIN)       "
164 HDR$ = HDR$ + "FREQ    F(COS)         F(SIN)": PRINT : PRINT
166 GOSUB 100
168 RETURN

169 ' *** INVERSE TRANSFORM ***
170 SK1 = 1: K6 = 1: YSF = 64
172 CLS : HDR$ = "TIME     AMPLITUDE    NOT USED     "
174 HDR$ = HDR$ + "TIME    AMPLITUDE     NOT USED": PRINT : PRINT
176 GOSUB 100
178 RETURN

'   *****  OUTPUT DATA  *****
200 T9 = TIMER - T9
202 IF IOFLG = 1 THEN 250
206 ZSTP = 15
208 PRINT HDR$: PRINT : PRINT
210 FOR Z = 0 TO Q2 - 1
215 GOSUB 300
216 IF Z > ZSTP THEN 350 ' PRINT 1 SCREEN AT A TIME
220 NEXT Z
222 LOCATE 1, 1: PRINT : PRINT "TIME ="; T9
225 INPUT "C/R TO CONTINUE:"; A$
229 RETURN

250 ' *** PLOT OUTPUT ***
252 CLS : YPX = 0: IF K6 = -1 THEN 280
254 Y0 = 175
256 FOR I = 0 TO Q - 1
258 IF C(I) > YPX THEN YPX = C(I)
260 NEXT I
262 YSF = 100 / YPX
263 LINE (X0 + 10, Y0 - 100)-(X0, Y0 - 100)
264 LOCATE 6, 1: PRINT USING "###.##"; YPX
265 LINE (X0 - 1, 50)-(X0 - 1, 300)
266 LINE (X0, Y0)-(X0 + 500, Y0)
268 LINE (X0, Y0)-(X0, Y0)
```

Transient Analysis

```
270 FOR I = 0 TO Q - 1
272 YP = C(I)
274 IF K6 = -1 THEN YP = SQR(C(I) ^ 2 + S(I) ^ 2)
276 LINE -(X0 + XSF * I, Y0 - YSF * YP)
278 NEXT I
279 GOTO 222
280 ' ***   FIND Y SCALE FACTOR FOR FREQ DOMAIN PLOT   ***
281 Y0 = 300 ' SET Y AXIS ORIGIN
282 FOR I = 0 TO Q - 1 ' FIND LARGEST VALUE IN ARRAY
284 YP = SQR(C(I) ^ 2 + S(I) ^ 2): IF YP > YPX THEN YPX = YP
286 NEXT I
287 YSF = 200 / YPX ' SET SCALE FACTOR
288 LINE (X0 + 10, Y0 - 200)-(X0, Y0 - 200)
289 LOCATE 8, 1: PRINT USING "###.##"; YPX: GOTO 265

298 '***   PRINT DATA TO CRT   ***
300 PRINT USING "###"; Z; : PRINT "   ";
310 PRINT USING "+##.#####"; C(Z); : PRINT "   ";
312 PRINT USING "+##.#####"; S(Z); : PRINT "   ";
320 PRINT USING "###"; Z + Q2; : PRINT "   ";
322 PRINT USING "+##.#####"; C(Z + Q2); : PRINT "   ";
324 PRINT USING "+##.#####"; S(Z + Q2)
330 RETURN

350 ' *** STOP WHEN SCREEN FULL ***
352 ZSTP = ZSTP + 16
354 PRINT : INPUT "C/R TO CONTINUE:"; A$
356 CLS : PRINT HDR$: PRINT : PRINT
358 GOTO 220

600 ' *** SQUARE FUNCTION ***
602 CLS : PRINT : PRINT
604 PRINT "PREPARING DATA INPUT - PLEASE WAIT!"
610 FOR I = 0 TO Q / 2 - 1
620 C(I) = -1: S(I) = 0
630 NEXT
640 FOR I = Q / 2 TO Q - 1
650 C(I) = 1: S(I) = 0
660 NEXT
680 GOSUB 160 ' TRANSFORM TO FREQUENCY DOMAIN
690 RETURN

699 ' ***   MODIFY SYSTEM PARAMETERS   ***
700 CLS
702 INPUT "PRESENT DATA GRAPHICALLY? (Y/N):"; A$
704 IF A$ = "Y" OR A$ = "y" THEN IOFLG = 1 ELSE IOFLG = -1
706 RETURN

'       ****************************************
990 STOP: END
'       ****************************************
```

```
1000 ' ****************************************
1002 ' *            MODIFY SPECTRUM           *
1004 ' ****************************************
1040 CLS : RTFLG = 0
1042 IF K6 <> -1 THEN 1096
1050 PRINT SPC(30); "MODIFY SPECTRUM MENU": PRINT : PRINT
1060 PRINT SPC(5); "1 = TRUNCATE SPECTRUM": PRINT
1062 PRINT SPC(5); "2 = BUTTERWORTH RESPONSE": PRINT
1064 PRINT SPC(5); "3 = GAUSSIAN RESPONSE": PRINT
1068 PRINT SPC(5); "6 = EXIT": PRINT
1070 PRINT SPC(10); "MAKE SELECTION :";
1080 A$ = INKEY$: IF A$ = "" THEN 1080
1082 A = VAL(A$): ON A GOSUB 1100, 1200, 1300, 1990, 1990, 1990
1084 GOSUB 200
1088 IF RTFLG = 1 THEN RETURN
1090 GOTO 1040
1096 INPUT "MUST TRANSFORM SIGNAL BEFORE MODIFYING SPECTRUM"; A$
1098 RETURN

1100 ' *** TRUNCATE SPECTRUM ***
1102 CLS : PRINT : PRINT
1104 INPUT "ENTER CUTOFF FREQUENCY"; A$
1110 FCO = VAL(A$): IF FCO = 0 OR FCO > Q3 THEN 1104
1130 FOR I = FCO TO Q - FCO
1140 C(I) = 0: S(I) = 0
1150 NEXT I
1194 RTFLG = 1: RETURN

1200 REM * BUTTERWORTH RESPONSE *
1202 CLS : PRINT : PRINT
1204 PRINT "ENTER THE 3db CUTOFF HARMONIC NUMBER (1 TO "; Q3; ")";
1206 INPUT A$
1208 FCO = VAL(A$): IF FCO = 0 OR FCO > Q3 THEN 1204
1210 NP = 7: REM NUMBER OF POLES = NP
1220 FOR I = 1 TO Q3
1222 ATTN = 1 / SQR(1 + (I / FCO) ^ (2 * NP))
1224 C(I) = ATTN * C(I): S(I) = ATTN * S(I)
1226 C(Q - I) = ATTN * C(Q - I): S(Q - I) = ATTN * S(Q - I)
1228 NEXT I
1230 ATTN = 1 / SQR(1 + (Q2 / FCO) ^ (2 * NP))
1234 C(Q2) = ATTN * C(Q2)
1294 RTFLG = 1: RETURN

1300 REM * GAUSSIAN RESPONSE *
1302 CLS : PRINT : PRINT
1304 PRINT "ENTER THE 3db CUTOFF HARMONIC NUMBER (1 TO "; Q3; ")";
1306 INPUT A$
1308 FCO = VAL(A$): IF FCO = 0 OR FCO > Q3 THEN 1304
1320 FOR I = 1 TO Q3
1322 ATTN = 1 / SQR(EXP(.3 * ((I / FCO) ^ 2)))
1324 C(I) = ATTN * C(I): S(I) = ATTN * S(I)
```

Transient Analysis 55

```
1326 C(Q - I) = ATTN * C(Q - I): S(Q - I) = ATTN * S(Q - I)
1328 NEXT I
1330 ATTN = 1 / SQR(EXP(.3 * ((Q2 / FC0) ^ 2)))
1334 C(Q2) = ATTN * C(Q2): S(Q2) = ATTN * S(Q2)
1394 RTFLG = 1: RETURN

1990 ' **** EXIT MODIFY SPECTRUM ROUTINE ****
1992 RTFLG = 1: RETURN
```

The core program (up to line 200) is our conventional FFT program (see Chapters 8 and 9 of *Understanding the FFT*); therefore, this description will cover only the subroutines incorporated to illustrate Gibb's phenomenon.

The *generate function* routine starts at line 600. At line 610 we set up a loop to insert -1 into the *real* data array (i.e., C(I)), and clear the *imaginary array* (i.e., set S(I) = 0 clearing any data left over from previous operations). This loop steps from 0 to Q/2 -1 (i.e., the first half of the array). At line 640 we do the same thing for the second half of the array except that we set C(I) to +1. We then jump down to the forward transform routine (line 680), and when we come back to line 690, the frequency domain function will be ready for our experiments.

```
600 ' *** SQUARE FUNCTION ***
602 CLS : PRINT : PRINT
604 PRINT "PREPARING DATA INPUT - PLEASE WAIT!"
610 FOR I = 0 TO Q / 2 - 1
620 C(I) = -1: S(I) = 0
630 NEXT
640 FOR I = Q / 2 TO Q - 1
650 C(I) = 1: S(I) = 0
660 NEXT
680 GOSUB 160 ' TRANSFORM TO FREQUENCY DOMAIN
690 RETURN
```

The Modify Spectrum routine starts at line 1000. We first clear the screen and the Return Flag. At line 1042 we check the K6 flag to see if a forward transform has just been made (i.e., to see if a non-modified spectrum is available—if not, we let the user know at line 1096). We display the menu of options available to the user (lines 1050 - 1080) and, at the user's option, jump to the appropriate subroutine. Note that there are only six selections, but the ON/GOSUB instruction allows A to be anything between 0 and 255. If you push the wrong key the menu will just recycle because RTFLG will still equal zero (line 1088 and 1090).

```
1000 ' ****************************************
1002 ' *            MODIFY SPECTRUM           *
1004 ' ****************************************
1040 CLS : RTFLG = 0
1042 IF K6 <> -1 THEN 1096
1050 PRINT SPC(30); "MODIFY SPECTRUM MENU": PRINT : PRINT
1060 PRINT SPC(5); "1 = TRUNCATE SPECTRUM": PRINT
1062 PRINT SPC(5); "2 = BUTTERWORTH RESPONSE": PRINT
1064 PRINT SPC(5); "3 = GAUSSIAN RESPONSE": PRINT
1068 PRINT SPC(5); "6 = EXIT": PRINT
1070 PRINT SPC(10); "MAKE SELECTION :";
1080 A$ = INKEY$: IF A$ = "" THEN 1080
1082 A = VAL(A$): ON A GOSUB 1100, 1200, 1300, 1990, 1990, 1990
1084 IF RTNFLG = 1 THEN GOSUB 200: RETURN' PLOT DATA AND EXIT
1090 GOTO 1040
1096 INPUT "MUST TRANSFORM SIGNAL BEFORE MODIFYING SPECTRUM"; A$
1098 RETURN
```

If you select option *1 = Truncate Spectrum* from the *Modify Spectrum Menu*, the jump is to line 1100. Line 1104 asks for the *cutoff frequency*, above which all harmonics will be eliminated. If you have selected 2^{10} data points (i.e., Q = 1024) then FCO *max* will obviously be 512 which is verified in line 1110. Line 1130 sets up a loop to remove all harmonics above the selected cutoff frequency. Note that we remove the *negative frequencies* from the negative frequency part of the spectrum (i.e., from FCO to Q - FCO). When we have accomplished this we set RTFLG to 1 and the job is done.

```
1100 REM * TRUNCATE SPECTRUM *
1102 CLS : PRINT : PRINT
1104 INPUT "ENTER CUTOFF FREQUENCY"; A$
1110 FCO = VAL(A$): IF FCO = 0 OR FCO > Q3 THEN 1104
1130 FOR I = FCO TO Q - FCO
1140 C(I) = 0: S(I) = 0
1150 NEXT I
1194 RTFLG = 1: RETURN
```

Type this program into your BASIC application software and run it. The FFT can handle reasonably large transforms (in a matter of seconds) so select 2^{10} components to provide plenty of resolution. First, simply transform the step function and reconstruct (to verify the original time domain function is obtained). The results will be as shown in Figure 6.3 below—note that Gibbs' phenomenon does not appear when we do not modify the spectrum. Next, repeat the transform and select the *MODIFY SPECTRUM* routine. Remove a large portion of the spectrum (e.g., from

Transient Analysis

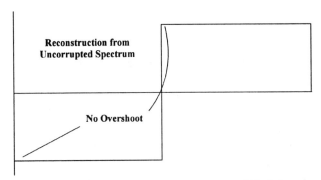

Figure 6.3 - Reconstruction from Unmodified Spectrum

the 60th harmonic up). Reconstruct the time domain signal and you will observe that Gibbs' phenomenon is present. Visually, things may not be very clear; nonetheless, Gibbs' phenomenon *is* apparent.

Repeat the illustration, but this time remove everything from the 16th harmonic up. Now when we reconstruct the time domain the oscillations at the transitions are readily apparent (Figure 6.4). So, Gibbs' phenomenon does not appear *simply* because we transform and reconstruct

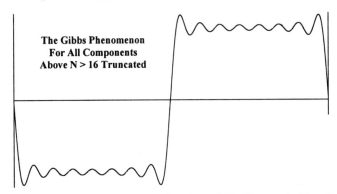

Figure 6.4 = Reconstruction from 16th Harmonic Cutoff

the data; however, when we drop off the higher frequency components, it definitely does appear. *[Did you notice anything peculiar? We will have to come back to this. There are a couple of things in the above illustration that aren't completely on the up-and-up, but they have a perfectly good reason for being there, and we'll discuss them in detail shortly.]*

6.5 FILTERS

As we noted in the last chapter of *Understanding the FFT*, the response of a filters in the time domain, being the result of a convolution integral, isn't something we would expect to understand intuitively; however, in the frequency domain, filters are dead simple. Consider this example: a human can hear frequencies up to about 20 kHz (20,000 cycles/second). Now, suppose the speaker we use has a strong resonance above 20 kHz, and signals there may drive the speaker beyond its limits, causing *audible* signals to become distorted. We can eliminate this possibility by filtering off *all* harmonics above 20 kHz. If we remove the harmonics *above* 20 kHz from our audio signal few listeners will hear the difference. On the other hand, we want to reproduce all the frequencies *below* 20 kHz (i.e., the part we *do* hear) with high fidelity. This, then, is the basic idea of a filter—we want to remove certain harmonics from the signal (the *reject band*) while leaving all harmonics within another band of frequencies (the *pass band*) unchanged.

As it turns out, while we can do pretty amazing things with modern filters, we can't just truncate all frequencies above a certain cutoff (fortunately, most of the time that isn't necessary). What we *can* do is attenuate some frequencies with a filter circuit while passing others *essentially* unchanged. If we can accept a response that *approaches* the desired result, as opposed to a *perfect* solution, we can design circuits that do *almost* anything (ah, if we only knew what we were wishing for).

Now, if we know the *frequency response* (i.e., the transfer function—the *attenuation and phase shift*) of our filter, we can easily figure what will happen to the spectrum of *any input signal*. We simply multiply every harmonic of the input signal by the attenuation and phase shift of the circuit at that frequency, and we will have the frequency domain representation of the *output function*. We could then take the inverse transform of this output function and obtain the time response of the circuit.[3] [In the past, taking transforms and inverse transforms was such a painful task that engineers seldom bothered—it really wasn't necessary! *Most of the time* a knowledge of the transfer function alone (to the experienced eye) was *good enough*. The transfer function was, for the most part, sufficient to evaluate a filter circuit.]

---

[3]Much of this was covered in the last chapter of *Understanding the FFT*.

6.6 FILTERS—NOTIONS AND TERMINOLOGY

We will *not* discuss filter design here, but we *will* need to consider the system level performance of filters. We note that, with physically real circuits, it's impossible to filter a signal and not shift the phase (i.e., give different phase shifts to different frequencies). It's also impossible to leave the passband perfectly flat while completely removing frequencies beyond cutoff with real circuits (as we just did in the Gibbs illustration). What *can* we do with physically real filters then?

Let's use a real circuit example to get some terminology (and a few notions) straight. A very simple filter is provided by the RC network shown below. We recognize that the resistor R_1 and capacitor C_1 form a voltage divider. As frequency increases the reactance of the capacitor decreases causing more of the input voltage to be dropped across R_1 and less across C_1. Our objective is to remove frequency components above

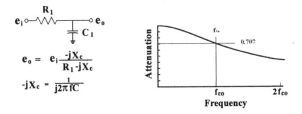

Figure 6.5 - RC Filter & Frequency Response

the cutoff frequency, f_{co}, while leaving harmonics below f_{co} unchanged. For physically realizable filters, however, the amplitude *rolls-off* over some range of frequencies. We must therefore decide where *cutoff* takes place—where the *pass* band ends and the *reject* band begins. We might select the frequency where the capacitive reactance equals the resistance as a sort of half-way point; however, because of the *quadrature* relationship between an *imaginary* reactance and a *real* resistance, this attenuates the signal to $1/\sqrt{2} = 0.707$. At this attenuation the *power* delivered to a resistive load is attenuated to half ($P = V^2/R$). Equating resistance to reactance doesn't apply in more complex filters, but the notion of *half-power* will always apply; so, the half-power point becomes the universal definition of *cut-off*. Furthermore, half-power equals 3.0103 deci-bels (db) attenuation, so we refer to the cut-off as the *3 db* point.

But the circuit of Fig. 6.5 is not a very good filter. We might improve things by connecting several RC sections end-to-end, with each section sort of repeat-filtering the output of the preceding section; but, while the notion of increasing the number of sections (i.e., number of *poles*[4]) is indeed valid, cascading RC sections will never produce a truly high performance filter. *High performance* filters require both inductors and capacitors in each section (or active components—i.e., op amps).

6.7 HIGH PERFORMANCE FILTERS

There are three major design criteria for filters: Chebychev, Butterworth and Gaussian (there are other design criteria however). The Chebychev and Butterworth are essentially the same class of filter, with the Chebychev trading flatness of the passband for steepness of rolloff, while the special case of Butterworth tries to achieve the idealized objective we put forth above (i.e., keeping passband flat—see fig. below). [We will only consider the Butterworth here as it illustrates our present concerns adequately.]

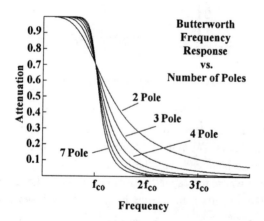

Figure 6.6 - Butterworth Filter Response

---

[4] For the filters we will be considering here, the number of *poles* will be equal to the number of reactive elements in the circuit. The *Transfer Function* of a circuit may be expressed as the ratio of two frequency domain polynomials. The roots of a polynomial are the values of the independent variable for which the function goes to zero. The roots of the numerator are the *zeros* of the transfer function while the roots of the denominator are *poles*—'nuff said.

Transient Analysis

In our *Gibbs illustration* program we have included a subroutine which emulates the performance of a Butterworth filter. If (under the modify spectrum menu) we select *2 = Butterworth Response* we will go to line 1200. This modification is a little more subtle although we are still attenuating the frequencies—similar to the previous routine. Again we input in the cutoff frequency (i.e., the 3db point). We do not ask how many poles you want the filter to be—we simply set NP = 7 in line 1210. (Note: we use the 7-pole configuration since it comes closest to what we desire in Figure 6.6 above—you may change this if you like). We then calculate the attenuation for this number of poles and attenuate the positive and negative spectrums (i.e., both up and down to Q3). Starting at line 1220 we set up a loop which steps through the frequency components, calculates the attenuation, and multiplies both positive and negative frequencies by this factor (lines 1222 - 1228). We then solve for the attenuation of the "Nyquist frequency" in a separate step by attenuating the cosine component (there's no sine component), set RTFLG = 1, and exit.

```
1200 REM * BUTTERWORTH RESPONSE *
1202 CLS : PRINT : PRINT
1204 PRINT "ENTER THE 3db CUTOFF HARMONIC NUMBER (1 TO "; Q3; ")";
1206 INPUT A$
1208 FCO = VAL(A$): IF FCO = 0 OR FCO > Q3 THEN 1204
1210 NP = 7: REM NUMBER OF POLES = NP
1220 FOR I = 1 TO Q3
1222 ATTN = 1 / SQR(1 + (I / FCO) ^ (2 * NP))
1224 C(T1, I) = ATTN * C(T1, I): S(T1, I) = ATTN * S(T1, I)
1226 C(T1, Q-I) = ATTN * C(T1, Q-I): S(T1, Q-I) = ATTN * S(T1, Q-I)
1228 NEXT I
1230 ATTN = 1 / SQR(1 + (Q2 / FCO) ^ (2 * NP))
1234 C(T1, Q2) = ATTN * C(T1, Q2)
1294 RTFLG = 1: RETURN
```

Load FFT06-01 again, run the Butterworth subroutine, and then reconstruct the time domain waveshape. As you will see, Gibbs' phenomenon is indeed present.

Let's take the time to note that Gibbs is not a trivial phenomenon. While it may be argued that this does *not* represent distortion (the harmonics were there in the original input signal), for many applications overshoot and ringing are simply unacceptable.

It begins to appear our earlier exposition of this phenomenon is valid—if we remove the upper harmonic components we get Gibbs' overshoot in the time domain; but there's something curious in this result.

We know all real systems are band limited, but we also know that overshoot and ringing does *not* appear on *every* signal that has a rapid transition. Why not? Obviously we still don't have the whole story here.

6.8 GAUSSIAN FILTERS

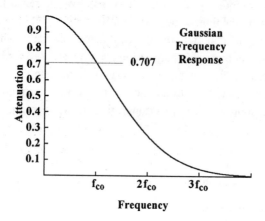

Fig. 6.7 - Gaussian Response

There *is* a way to avoid Gibbs' phenomenon of course. If, instead of simply *lopping-off* the harmonics above cutoff, we gradually decrease their amplitude, we can fool the waveshape into *not knowing* what it's *missing*. The Gaussian response accomplishes this, but the shape of the attenuation curve sacrifices both flat passband *and* sharp *knee* at the cutoff frequency (Fig. 6.7). Unfortunately, this can't be improved by adding more poles to the filter—this *is* the shape required to eliminate overshoot.[5] If we add more poles we must do it in such a way that the gradual rolloff is not changed—the shape of the attenuation curve *must* be preserved.

As with Butterworth, we have included a routine in FFT06-01 to emulate a Gaussian response. If you select *3 = Gaussian response* from the Modify Spectrum Menu, you will go to line 1300. Once again (in line 1304) we select the 3db cutoff frequency. At line 1320 we set up an identical loop to the Butterworth routine except that we solve a different equation for attenuation.

---

[5]There is a class of filter (Optimum Transient) which allows some overshoot but minimizes the time to settle to steady state value (within some accuracy).

Transient Analysis

```
1300 REM * GAUSSIAN RESPONSE *
1302 CLS : PRINT : PRINT
1304 PRINT "ENTER THE 3db CUTOFF HARMONIC NUMBER (1 TO "; Q3; ")";
1306 INPUT A$
1308 FCO = VAL(A$): IF FCO = 0 OR FCO > Q3 THEN 1304
1320 FOR I = 1 TO Q3
1322 ATTN = 1 / SQR(EXP(.3 * ((I / FCO) ^ 2)))
1324 C(T1, I) = ATTN * C(T1, I): S(T1, I) = ATTN * S(T1, I)
1326 C(T1, Q-I) = ATTN * C(T1, Q-I): S(T1, Q-I) = ATTN * S(T1, Q-I)
1328 NEXT I
1330 ATTN = 1 / SQR(EXP(.3 * ((Q2 / FCO) ^ 2)))
1334 C(T1, Q2) = ATTN * C(T1, Q2): S(T1, Q2) = ATTN * S(T1, Q2)
1394 RTFLG = 1: RETURN
```

You may run this Gaussian response routine to verify that overshoot does not appear. You still don't get to select the number of poles—this attenuation curve is for an infinite number of poles (we spare no expense); however, except for frequencies well beyond cutoff, the response doesn't change much vs. number of poles.

6.9 TRANSIENT RESPONSE

When we filter the waveform via any of the above routines we display the modified spectrum before jumping back to the Main Menu. The displayed spectrum *must* be the same as a signal that has been filtered by a circuit with the selected *frequency response*. In the above examples, it must be the response that would be obtained if we passed a square wave through a circuit with the selected frequency response! This, in fact, is the key concept we want to convey here: the transient response of a circuit (for any input waveform we choose) is found by multiplying the *transform* of that waveform by the frequency response of the circuit. That is, by convolution of the time domain functions.

Now, when we found the transient response of a Butterworth filter above, we simply multiplied the spectrum of a step function by an *equation* describing the frequency response of such a filter; however, in the general case, finding equations for the frequency response of arbitrary circuits is not an easy task. We need not trouble ourselves with trying to find such equations, of course, because we can find the frequency response of any circuit via the network analysis program of the previous chapter.

Using the circuit analysis program developed in Chapter 5 we may determine the attenuation and phase characteristics of Butterworth filters (or any circuit). Once we obtain this *transfer function*, we may use it to modify the spectrum of a square wave (or any other wave). This requires nothing more than multiplication of the spectrum of the square wave by the frequency response of the filter (complex multiplication of course). By simply performing the inverse transform then, we obtain the transient response of the circuit.

The two programs (i.e., ACNET02-01 and FFT06-01) are combined or *merged* in Appendix 6.1 to obtain a program which provides the transient response of a circuit. The program will work pretty much as the Gibbs Phenomenon program did, except that now we can *enter a circuit* (from the Modify Spectrum Menu) and multiply the circuit transfer function with the spectrum of the step function. Just as before, we find the transient response via an inverse transform of the results. [Note: While the program given is acceptable for the purpose of illustration, you might want to expand it so that it can perform either frequency response analysis or transient analysis. It's not such a difficult task, and on today's high speed personal computers the execution time is not unacceptable (especially with compiled BASIC). Even if you have access to *heavyweight* circuit analysis programs, you might still find this application handy on occasions. With only a modest effort you can add *dependent generators*, which will allow inclusion of active components.

Before we leave this section, let's consider one of the peculiar aspects of our Gibbs phenomenon discussion. When we considered Butterworth filters, for example, we found the ringing and overshoot started *before* the transition! Now, that's not how it should happen in the real world— if overshoot and ringing are *caused* by a rapid transition they should occur *following* the driving transient.... What's wrong here?

The reason for the prescient ringing and overshoot in the above is that we used only the amplitude response of the filter and ignored the phase. The reason we ignored the phase response was that we had a simple but accurate equation for amplitude response and it was easy to modify the spectrum by this equation.[6] There's a better way, of course. When we determine the transient response via circuit analysis, we automatically include the phase response, and the result gives a much more truthful picture of the transient response.

---

[6] There's another reason for using only the amplitude response—it's sometimes desirable to filter signals without disturbing phase, and we have illustrated how the FFT allows that operation. This technique might come in handy some day.

Transient Analysis

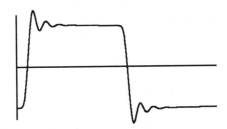

Fig. 6.8 - Transient Response of Butterworth Filter

6.10 ALIASING

The above takes care of the *funny* transients caused by ignoring phase, but there was another anomalous condition that existed in the preceding exposition. We generated the input square wave by simply placing "+1"s and "-1"s in the input buffer, and we *know* we can't do that. If we digitize a signal with frequencies above the Nyquist we *know* we will get *aliasing*. We will discuss aliasing in greater detail later but, for now, we can avoid the problem of aliasing if we simply *generate* the frequency domain function of a square wave in the first place. This is only slightly different from what we have been doing when we generate the *time domain* function by summing the contributions from each harmonic—we can generate the frequency domain function as follows:

```
600 REM * FREQUENCY DOMAIN SQUARE WAVE FUNCTION *
602 CLS : PRINT : PRINT
604 PRINT "PREPARING DATA INPUT - PLEASE WAIT!"
606 FOR I = 0 TO Q: C(I) = 0: S(I) = 0: NEXT
610 FOR I = 1 TO Q3 STEP 2
612 S(I) = -.6366196 / I:S(Q - I) = .6366196/ I
614 NEXT
622 SK1 = 2: K6 = -1: YSF = Q
628 RETURN
```

If you generate this input and then immediately perform the inverse transform, you will find Gibbs' phenomenon prominently displayed at the corners—*not* because we have taken the transform, but simply because we have started out with the *incomplete*, band limited, square wave spectrum.

Note that, even though our first input signal was "corrupted" by deliberate aliasing, the inverse transform still produced the "correct" *waveshape—which is interesting. The FFT knows nothing about aliasing—it simply does its job and finds the transform of the data. We know, however, the spectrum obtained was* not *the same as a real square wave.*[7]

CIRCUIT ANALYSIS vs FOURIER

Finally, we should note the *similarity* between network analysis and Fourier analysis. What we are doing in circuit analysis is solving the transfer impedance equations for a band of unit amplitude sinusoids (i.e., at a lot of different frequencies). The result, of course, is simply the transfer function, in the same format as the DFT transform of a function. We shall see later that a spectrum of equal amplitude sinusoids may be identified as an impulse function, and if we take the inverse transform of the frequency response itself, we will see the transient response of the circuit to an impulse.

So, when we find the transient response of any circuit as we have done above, we are simply convolving the impulse response of the circuit with the transform of a square wave (or other function of our choosing). While this may seem like a *round-about* way of doing things, it's a very practical method of solving real world problems...but for us it's primarily another step toward understanding Fourier analysis.

---

[7] If we look at what the summation of these harmonics produce in the *continuous* world, we find that it doesn't really produce a "flat-topped" time domain function either. We will investigate the technique of *looking between the data points* a little later.

CHAPTER 7

SPEEDING UP THE FFT

7.0 INTRODUCTION

It's well known there are faster FFT algorithms than the classic Cooley-Tukey derivation. "Tricks" that may be used to speed up the transform become apparent when we analyze the FFT *mechanics*. For example, in the first two stages, where we are dealing with one and two-point DFTs, the twiddle factors have values of +1, -1, and zero, and we need not evaluate a Taylor series for these—nor perform a multiplication at all. In fact, these first stages yield no sine components at all. We can save time here by writing simplified routines for the first two stages of computation and then revert to the conventional algorithm for the remainder. The modification will look something like this:

```
104 REM     *** STAGE A ***
106 FOR I = 0 TO Q2 - 1
108 CTEMP1 = C(I) + C(Q2 + I)
110 C(Q2 + I) = C(I) - C(Q2 + I): C(I) = CTEMP1
112 NEXT I
114 REM     *** STAGE B ***
116 Q34 = Q2 + Q4  ' 3/4 ARRAY LOCATION
118 FOR I = 0 TO Q4 - 1: IQ2 = I + Q2: IQ4 = I + Q4: IQ34 = I + Q34
120 CTEMP = C(I) + C(IQ4): C(IQ4) = C(I) - C(IQ4): C(I) = CTEMP
122 S(IQ2) = C(IQ34): S(IQ34) = -C(IQ34): C(IQ34) = C(IQ2)
124 NEXT I
```

We then enter the conventional FFT routine starting at $M = 2$:

```
132 FOR M=2 TO N-1 : .....
```

[Note: The validity of this *trick* may not yet be apparent but we will consider it in more detail later (under *small array* FFTs)—for now I recommend you try this modification in the FFT routine of the previous chapter; however, keep in mind that this modification works *only for the forward transform.*]

This will cut off a little processing time and, as we said, every little bit helps; however, there are other, much more significant, ways to speed things up.

7.1 ELIMINATING NEGATIVE FREQUENCY COMPONENTS

As you will recall, the development of the FFT was based on a peculiarity of a *stretched data* configuration wherein half of the frequency components were redundant. That is, the spectrum of a stretched function is a *double image* (i.e., the upper half of the spectrum duplicates the lower). This redundancy made it possible to eliminate the work of extracting the upper half frequency components. By pursuing this phenomenon we eventually wound up with an algorithm for the FFT.

Now, we could *not* have developed an FFT algorithm without the intrinsic symmetry and redundancy entrained in the conventional DFT. Furthermore, it's apparent we haven't removed *all* of the redundancy, for as we have noted, when we perform a DFT (or, for that matter, an FFT) on real data the negative frequencies (i.e., those above the Nyquist frequency) are identical to the positive harmonics, except for negation of the sine term (i.e., they are the complex conjugates of the positive frequencies). Why, then, should we go to the trouble of calculating these components? We can simply find the positive frequency terms and repeat them in the data printout (negating the sine components of course).

But then, why should we bother with them at all? Why modify redundant frequency components when we, for example, filter a signal?... or differentiate a function?...or perform any other operation in the frequency domain? It would seem likely that we can cut our work in half (very nearly) if we simply *eliminate* the negative frequency terms!

Okay, how must we modify our *conventional* FFT algorithm so we can eliminate the time required to determine negative frequencies?

7.2 STRETCHING THEOREM (Positive Frequencies Only)

In the DFT we may easily eliminate the negative frequency harmonics by simply ending our procedure when we reach the Nyquist frequency, but how will this affect the FFT algorithm? If we ignore the negative frequencies, what happens to the positive frequencies when we transform stretched data? [Our purpose may not be completely clear here, but we are simply trying to illustrate the mechanics of a *positive frequency only* transform—the significance of this will be apparent shortly.]

We borrow a program used in Chapter 5 of *Understanding the FFT* which illustrates the stretching theorem. This program is not an FFT,

The PFFFT

but rather, a DFT—we modify it here to yield only positive frequencies (see listing in Appendix 7.1). This program yields the following spectrum for the *un-stretched* function:

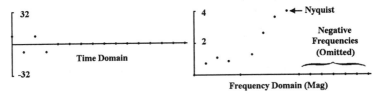

Figure 7.1 - Transform with Positive Frequencies Only

We have obviously eliminated the upper half of the spectrum—the negative frequencies are gone—we didn't compute them (saving ourselves a great deal of work). But *what happens when we stretch this function?* Fig. 7.2 below shows the effect of stretching (still using a positive frequency transform). We must look closely at this result. We *are* getting a spectrum duplication but it's no longer a simple *spectrum doubling*.

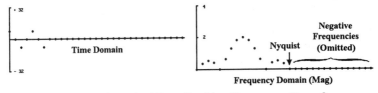

Figure 7.2 - Stretched Data Positive Frequency Transform

Stretching the function when only the positive frequencies are considered creates the *mirror image* of the positive frequency components (still in the positive frequency *space*). While this is *not* the same thing that was obtained when we stretched a function using the conventional DFT (i.e., with both positive and negative frequency components), the underlying mechanism *is* the same. The subtle difference is important for us—it defines the *positive frequency only* FFT.

We need to understand what's happening here. It's as if we have introduced a *secondary sampling pulse train* which is half the real sampling rate. Data sampling is like amplitude modulation—the data *modulates* the sampling pulses.[1] From this point of view, the effect of stretching the data is as if we are modulating a pulse train that is half the

---

[1] Amplitude modulation of a "carrier" (e.g., unit amplitude sampling pulses) produces "side bands" both above and below the carrier frequency. We will discuss data sampling and this modulation phenomenon in detail later.

real sampling rate. So, then, even though we are only finding the positive frequencies (relative to the real sampling rate), we will see the positive and negative side bands about this half frequency pseudo sampling rate. The unique thing here is that stretching a function (when only the positive frequencies are involved) creates a *mirror image* of the un-stretched spectrum *above* the pseudo sampling rate. This is the same mechanism (but *not quite* the same result) as the simple repeat effect we obtained in our original development of the FFT. [*Note: In both cases the negative frequency components are complex conjugates of the positive frequency components, but when you take the complex conjugate of a complex conjugate you get the original expression. If you are retired, with more time than pension, or if you are incurably curious, it may be worth your time to work this out—see* The Stretching Theorem, *Appendix 5.3, in* Understanding the FFT. *If you understand Figs. 7.1 and 7.2, however, that's all you need to know to create a Positive Frequency FFT.]*

7.3 THE PFFFT[2] (Positive Frequency Fast Fourier Transform)

Once again, we start with an 8 point data array *[You might want to read the FFT development given in* Understanding the FFT *before you begin this presentation]* :

$$\left| \begin{array}{c} DATA0 \\ ARRAY \end{array} \right| = |D_0, D_1, D_2, D_3, D_4, D_5, D_6, D_7| \quad \text{--------} \quad (7.0)$$

We separate the odd and even data elements creating two stretched arrays:

$$|DATA1'| = |D_0, 0, D_2, 0, D_4, 0, D_6, 0| \quad \text{--------} \quad (7.1)$$

$$|DATA2'| = |0, D_1, 0, D_3, 0, D_5, 0, D_7| \quad \text{--------} \quad (7.2)$$

The *positive frequency* transform of (7.1) will be:

$$\text{Transform} |DATA1'| = |F_0, F_1, F_2, F_1^*, F_0^*| \quad \text{--------} \quad (7.3)$$

(Where F_1^* and F_0^* are complex conjugates of F_1 and F_0)

We remove the zeros separating the data points in array $|DATA1'|$, and obtain the array:

$$|DATA1| = |D_0, D_2, D_4, D_6| \quad \text{--------} \quad (7.4)$$

---

[2] While I frown on mindless creation of acronyms, this mindless creation is an exception. The "P" is silent (as in swimming) yielding a slightly aspirated "ffft!"

The PFFFT

The positive frequency transform of this un-stretched array is:

Transform $|DATA1| = |F_0, F_1, F_2|$ -------------------- (7.5)

where F_0, F_1 and F_2 are the same in both Eqns. (7.3) and (7.5).

     Very well then, we must transform two arrays (i.e., the odd and even data points), and add the spectrums obtained. As before, we must rotate the phase of the *odd data transform* (due to the shifting phenomenon) before adding it to the *even data transform*. That is, the frequency components from the *odd data* transform are properly phase-shifted and

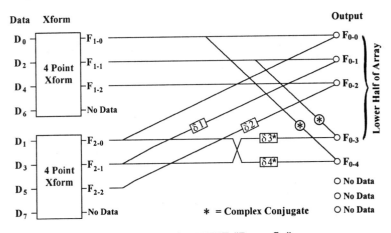

Figure 7.3 - The PFFFT "Butterfly"

summed with the frequency components of the *even data* transform; but, this time, the virtual components are created as the *mirror image* of the direct components (i.e., the complex conjugates are summed in reverse order). This is important so let's walk through the software for this algorithm just to make sure we understand how it really works.

     As before: Q = the number of data points.
                  K1 = 2*PI/Q
                  Q2 = Q/2;   Q4 = Q/4
                  J counts the frequency components.
                  I counts the data points.
                  Y(I) = input data
                  C(J) = cosine components of frequency
                  S(J) = sine components of frequency

```
109 REM * COMPUTE EVEN DFT *
110 FOR J=0 TO Q4:J1=K1*J
111 C(J) = 0: S(J)= 0
112 FOR I = 0 TO Q-1 STEP 2
114 C(J) = C(J) + Y(I)*COS(J1*I)
116 S(J) = S(J) + Y(I)*SIN(J1*I)
118 NEXT I
120 C(J)=C(J)/Q2:S(J)=S(J)/Q2
122 NEXT J
```

This routine is just a positive frequency DFT which operates on the *even* data points (the I counter counts in increments of 2). As you recall, we multiply the data points by sine and cosine functions (lines 114-116) and sum the products into the C(J) and S(J) arrays. The summations, of course, become the frequency components. We solve for the even data points here—the odd data transform will be stored in the upper half of the data arrays [i.e., C(J+Q2) and S(J+Q2)]. Note that in line 110 the J counter only counts up to Q/4 since we are only finding the positive frequencies (we have Q/2 data points since we are dealing with the even data points—therefore only Q/4+1 *positive* frequencies). We must, of course, take the DFT for the odd data points in a similar routine:

```
124 REM * COMPUTE ODD DFT *
126 FOR J=0 TO Q4:JO = J+Q2:J1=K1*J
127 C(JO) = 0:S(JO) = 0
128 FOR I=0 TO Q-1 STEP 2
130 C(JO)=C(JO)+Y(I+1)*COS(J1*I)
132 S(JO)=S(JO)+Y(I+1)*SIN(J1*I)
134 NEXT I
136 C(JO)=C(JO)/Q2:S(JO)=S(JO)/Q2
138 NEXT J
```

We have now taken the transform of each half of the data base in $(Q/4+1) * Q/2$ or $Q^2/8 + Q/2$, for a grand total of $Q^2/4 + Q$ (as opposed to Q^2 operations in the DFT or $Q^2/2$ for the conventional FFT). When Q is large this cuts processing time in half *again*...almost.

Okay, to obtain the complete transform we will sum the two partial transforms as follows:

```
139 ' *** SUM ODD AND EVEN TRANSFORMS (FINAL STAGE) ***
140 FOR J = 0 TO Q4: J1 = J+Q2: JI= Q2-J: K2 = K1*J: K3 = K1*JI
142 CTEMP1 = C(J) + C(J1)*COS(K2) - S(J1)*SIN(K2)
144 STEMP1 = S(J) + C(J1)*SIN(K2) + S(J1)*COS(K2)
146 CTEMP2 = C(J) + C(J1)*COS(K3) + S(J1)*SIN(K3)
148 S(JI) = C(J1)*SIN(K3) - S(J1)*COS(K3) - S(J)
150 C(J) = CTEMP1: S(J) = STEMP1: C(JI) = CTEMP2
152 NEXT J
```

The PFFFT

We simply sum the frequency components of the *even* DFT with the rotated components of the *odd* DFT. The virtual components are formed (lines 146-148) by summing the components in a mirror image (note that the index JI handles the virtual components in reverse order while storing the data in the location from which it came). For example, the first two components computed are the zero frequency and Nyquist components, both of which can be put back into the locations from which they came. If you step through this routine you will find the direct components may always be placed in the locations from which they came while the virtual components are always placed in empty locations. You may type this into your computer and verify that it works if you like.

7.4 EXTENDING THE CONCEPT

As with the conventional FFT, we extend this concept by splitting each *4-point DFT* into two *2-point DFTs*; however, before we actually discuss how to handle the PFFFT at the 2-point DFT level, let's take time to look at the incredible simplicity of these small array DFTs. In Figure 7.4 we show diagrammatically what takes place in these small DFTs. As

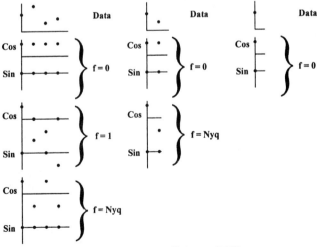

Figure 7.4 - Small Array DFTs

noted earlier, with four data points (or less) the sines and cosines only take on values of ± 1 and 0, greatly simplifying things.

Figure 7.4 shows a 4-point DFT (vertically at the left), a 2-point

DFT (running down the middle of the picture) and a 1-point DFT (right). We obtain the frequency components of the DFT by multiplying each data point by the corresponding point from the cosine and sine function (at the integer multiple frequencies—only positive frequencies are shown). It's obvious that, while we must compute the value of the sine and cosine for every data point in larger data arrays, calculating sines and cosines for these small arrays is hardly necessary (they are either +1, -1, or 0). This is the situation we took advantage of in the *speedup* technique discussed at the beginning of this chapter. [Note that, in Fig. 7.4, the 2-point DFT analyzes the *even* data points of the 4-point array (i.e., the 0 and 2nd data points). For the 8-point array we have been considering in this PFFFT illustration, the 4-point array would be the even data points of the whole array. If we were building a full FFT there would have to be a second DFT for the *odd* data points, of course (see Appendix 7.2).]

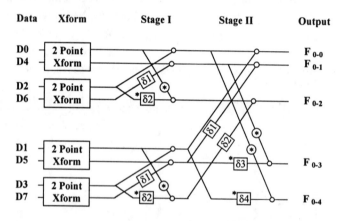

Figure 7.5 - PFFFT Double Butterfly

With this in mind we're ready to return to the development of the PFFFT. We illustrate the splitting of the 4-point DFT into two 2-point DFTs in Figure 7.5. Note that a 2-point DFT for positive frequencies is the same as a *conventional* DFT—both have a zero frequency and a Nyquist frequency. Now, within the 2-point DFT, the cosine term for the zero frequency component is just the sum of the zero and fourth data points divided by two (see Fig. 7.5). There is, of course, no sine term for the zero frequency nor the Nyquist component. The Nyquist (cosine) component is obtained by subtracting the fourth data point from the zero data point.

The PFFFT

For this illustration the 2-point DFTs will look like this:

```
110 C(0)=(Y(0)+Y(4))/2   '   PARTIAL XFORM
112 C(1)=(Y(0)-Y(4))/2   '   #0 STORED HERE
114 C(2)=(Y(2)+Y(6))/2   '   PARTIAL XFORM
116 C(3)=(Y(2)-Y(6))/2   '   #2 STORED HERE
118 C(4)=(Y(1)+Y(5))/2   '   PARTIAL XFORM
120 C(5)=(Y(1)-Y(5))/2   '   #1 STORED HERE
122 C(6)=(Y(3)+Y(7))/2   '   PARTIAL XFORM
124 C(7)=(Y(3)-Y(7))/2   '   #3 STORED HERE
```

When we sum these four partial transforms (to form the two 4-point transforms), our *positive frequency* transform (for 4 points) will only have *three* components. The phase shifts for the components from the *odd DFT* are still multiples of PI/2 (i.e., 90°) which yields:

```
126 CTMP = (C(0)+C(2))/2:S(0)=0   ' FOR f = 0 NO SINE
127 C(1) = C(1)/2:S(1)=C(3)/2
128 C(2) = (C(0)-C(4))/2:S(2)=0   ' FOR f = NYQ NO SINE
130 C(0) = CTMP
132 CTMP = (C(4)+C(6))/2:S(4)=0   ' NO SINE FOR f = 0
133 C(5) = C(5)/2:S(5)=C(7)/2
134 C(6) = (C(4)-C(6))/2:S(6)=0   ' NO SINE FOR f = NYQ
136 C(4) = CTMP
```

Note that we ignore the sine components when forming the next higher transform, and that we form only three positive frequency components. From here we may tack-on the final stage summation routine at the bottom of *p.*72 and obtain the complete transform.

Well and good but, carrying the FFT scheme to its logical conclusion, we must consider the transform subdivision process through one more step—until we are dealing with 1-point DFTs.

7.5 THE COMPLETE PFFFT

As illustrated in Figure 7.4, there is only one frequency component when there is only one data point. It will apparently be the zero frequency component, and is simply equal to the average value of the data point (i.e., equal to itself). Having noted this, it's apparent we may treat the input data points as single point transforms of themselves. In the first stage of computation, then, we will sum these single point transforms into 2-point transforms. Figure 7.6 (following page) shows the three stage summing process for the complete 8 data point PFFFT. It should be

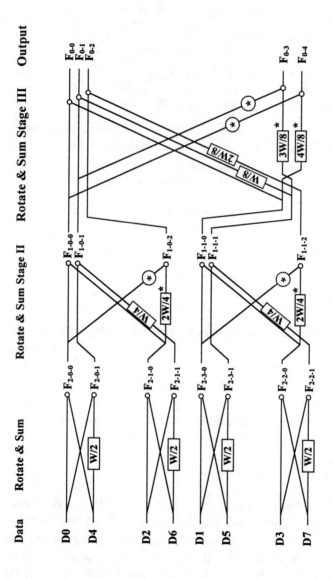

Figure 7.6 - Complete 8-Data Point PFFFT (Triple Butterfly)

The PFFFT

apparent that we are performing the same rotation and summing routine at every stage of computation, although it's also apparent the first two stages don't actually need the excess baggage of calculating sines and cosines (except to determine the *signs* of the summations). Let's look at the shifting and summing process for the first stage of computation: to form the zero frequency component of a 2-point transform we simply add the two data points required (see Fig. 7.6). Note that, since there's no sine component for a *zero* or *Nyquist* frequency component, the Nyquist frequency term is formed by negating the higher order data point and summing again (i.e., subtracting the two terms). For example, in lines 110 and 112 below, we do this for the 2-point transform of data points D0 and D4, and store them in locations C(0) and C(1).

```
109 '    ** SUM SINGLE POINT TRANSFORMS **
110 C(0) = (Y(0) + Y(4))/2'    PARTIAL XFORM
112 C(1) = (Y(0) - Y(4))/2'    #0 STORED HERE
114 C(2) = (Y(1) + Y(5))/2'    PARTIAL XFORM
116 C(3) = (Y(1) - Y(5))/2'    #1 STORED HERE
118 C(4) = (Y(2) + Y(6))/2'    PARTIAL XFORM
120 C(5) = (Y(2) - Y(6))/2'    #2 STORED HERE
122 C(6) = (Y(3) + Y(7))/2'    PARTIAL XFORM
124 C(7) = (Y(3) - Y(7))/2'    #3 STORED HERE
```

As with the conventional FFT then, we perform the PFFFT process by a shifting and summing mechanism from the beginning; however, we immediately see a problem here. In the above listed routine the partial xforms come out in sequence #0, #1, #2, and #3; but we want the partial transforms that are to be summed in the following stage to be stacked next to each other—as shown in the 2-point DFT routine at the top of *p*. 75. That is, if we hope to use a single rotate and sum routine for our PFFFT, we must manage the data such that we can easily determine data locations in each succeeding stage. In fact, if we examine the data locations in the above routine and the one on *p*. 75 (better yet, expand these to 16 or 32-point data arrays) we will see this is just our old *bit reversal* nemesis—knocking on the front door this time. We know how to handle bit reversal, of course, but now another alternative becomes apparent—it will be more efficient to bit-reverse the data in the first stage of the FFT (as opposed to *tacking-on* a *bit reversal* stage to the end). That is, we can take care of the *bit-reversal* directly in the first stage of computation by managing the array data such that all the partial transforms wind up where we want them. Here's how we will handle this:

```
101 ' ***  TRANSFORM STAGE 1  ***
110 C(0) = (S(0) + S(Q2))/2: C(1) = (S(0) - S(Q2))/2
112 FOR I = 1 TO Q3: I2 = 2*I: INDX = 0
114 FOR J = 0 TO N1: IF I AND 2^J THEN INDX = INDX + 2^(N - 2 - J)
116 NEXT J: INDX2 = INDX + Q2
118 C(I2) = (S(INDX) + S(INDX2))/2: C(I2+1) = (S(INDX) - S(INDX2))/2
120 NEXT I
122 FOR I = 0 TO Q1: S(I) = 0: NEXT I
```

Serendipity! We know the sine component in the first stage will be zero; so, we may put the input data into the sine component array S(J). We don't create the transforms in sequence, but bit-reverse an array index and thereby end up with the 2-point xforms in the desired sequence. From here we simply combine *adjacent* transforms to the end of the FFT.

7.6 BUILDING A GENERAL PURPOSE ALGORITHM

First we expand our illustration to 16 data points, then write separate loops for each partial transform:

```
100 '            *****   TRANSFORM   *****

101 ' ***  TRANSFORM STAGE 1  ***
102 C(0) = (S(0) + S(Q2))/2: C(1) = (S(0) - S(Q2))/2
104 FOR I = 1 TO Q3: I2 = 2*I: INDX = 0
106 FOR J = 0 TO N1: IF I AND 2^J THEN INDX = INDX + 2^(N - 2 - J)
108 NEXT J: INDX2 = INDX + Q2
110 C(I2) = (S(INDX) + S(INDX2))/2: C(I2 + 1) = (S(INDX) - S(INDX2))/2
112 NEXT I
114 FOR I = 0 TO Q1: S(I) = 0: NEXT I

' *********  SUM ODD/EVEN TRANSFORMS (FINAL STAGE -2)
116 FOR J = 0 TO 1
118 J1 = J + Q8: JI = Q8 - J: K2 = 4*K1*J: K3 = 4*K1*JI
120 CTEMP1 = C(J) + C(J1)*COS(K2) - S(J1)*SIN(K2)
122 STEMP1 = S(J) + C(J1)*SIN(K2) + S(J1)*COS(K2)
124 CTEMP2 = C(J) + C(J1)*COS(K3) + S(J1)*SIN(K3)
126 S(JI) = (C(J1)*SIN(K3) - S(J1)*COS(K3) - S(J))/2
128 C(J) = CTEMP1/2: S(J) = STEMP1/2: C(JI) = CTEMP2/2
130 NEXT J

132 FOR J = 0 TO 1: J0 = J + Q4
134 J1 = J + Q4 + Q8: JI = Q4 + Q8 - J: K2 = 4*K1*J: K3 = 4*K1*(JI - Q4)
136 CTEMP1 = C(J0) + C(J1)*COS(K2) - S(J1)*SIN(K2)
138 STEMP1 = S(J0) + C(J1)*SIN(K2) + S(J1)*COS(K2)
140 CTEMP2 = C(J0) + C(J1)*COS(K3) + S(J1)*SIN(K3)
142 S(JI) = (C(J1)*SIN(K3) - S(J1)*COS(K3) - S(J0))/2
144 C(J0) = CTEMP1/2: S(J0) = STEMP1/2: C(JI) = CTEMP2/2
146 NEXT J
```

The PFFFT

```
148 FOR J = 0 TO 1: J0 = J + Q2
150 J1 = J + Q2 + Q8: JI = Q2 + Q8 - J: K2 = 4*K1*J: K3 = 4*K1*(JI - Q2)
152 CTEMP1 = C(J0) + C(J1)*COS(K2) - S(J1)*SIN(K2)
154 STEMP1 = S(J0) + C(J1)*SIN(K2) + S(J1)*COS(K2)
156 CTEMP2 = C(J0) + C(J1)*COS(K3) + S(J1)*SIN(K3)
158 S(JI) = (C(J1)*SIN(K3) - S(J1)*COS(K3) - S(J0))/2
160 C(J0) = CTEMP1/2: S(J0) = STEMP1/2: C(JI) = CTEMP2/2
162 NEXT J

170 FOR J = 0 TO 1: J0 = J + Q34: J1 = J + Q34 + Q8
172 JI = Q34 + Q8 - J: K2 = 4*K1*J: K3 = 4*K1*(JI - Q34)
174 CTEMP1 = C(J0) + C(J1)*COS(K2) - S(J1)*SIN(K2)
176 STEMP1 = S(J0) + C(J1)*SIN(K2) + S(J1)*COS(K2)
178 CTEMP2 = C(J0) + C(J1)*COS(K3) + S(J1)*SIN(K3)
180 S(JI) = (C(J1) * SIN(K3) - S(J1) * COS(K3) - S(J0)) / 2
182 C(J0) = CTEMP1 / 2: S(J0) = STEMP1 / 2: C(JI) = CTEMP2 / 2
184 NEXT J

' *** SUM ODD AND EVEN TRANSFORMS (FINAL STAGE -1) ***
200 FOR J = 0 TO 2
201 J1 = J + Q4: JI = Q4 - J: K2 = 2*K1*J: K3 = 2*K1*JI
202 CTEMP1 = C(J) + C(J1)*COS(K2) - S(J1)*SIN(K2)
204 STEMP1 = S(J) + C(J1)*SIN(K2) + S(J1)*COS(K2)
206 CTEMP2 = C(J) + C(J1)*COS(K3) + S(J1)*SIN(K3)
208 S(JI) = (C(J1)*SIN(K3) - S(J1)*COS(K3) - S(J))/2
210 C(J) = CTEMP1 /2: S(J) = STEMP1 /2: C(JI) = CTEMP2/2
212 NEXT J
214 FOR J = 0 TO 2: J0 = J + Q2
156 J1 = J + Q34: JI = Q34 - J: K2 = 2*K1*J: K3 = 2*K1*(JI - Q2)
218 CTEMP1 = C(J0) + C(J1)*COS(K2) - S(J1)*SIN(K2)
220 STEMP1 = S(J0) + C(J1)*SIN(K2) + S(J1)*COS(K2)
222 CTEMP2 = C(J0) + C(J1)*COS(K3) + S(J1)*SIN(K3)
224 S(JI) = (C(J1)*SIN(K3) - S(J1)*COS(K3) - S(J0))/2
226 C(J0) = CTEMP1/2: S(J0) = STEMP1/2: C(JI) = CTEMP2/2
228 NEXT J

' *** SUM ODD AND EVEN TRANSFORMS (FINAL STAGE) ***
230 FOR J = 0 TO 4: J1 = J + Q2: JI = Q2 - J
232 CTEMP1 = C(J) + C(J1)*COS(K1*J) - S(J1)*SIN(K1*J)
234 STEMP1 = S(J) + C(J1)*SIN(K1*J) + S(J1)*COS(K1*J)
236 CTEMP2 = C(J) + C(J1)*COS(K1*JI) + S(J1)*SIN(K1*JI)
238 S(JI) = C(J1)*SIN(K1*JI) - S(J1)*COS(K1*JI) - S(J)
240 C(J) = CTEMP1: S(J) = STEMP1: C(JI) = CTEMP2
242 NEXT J
244 FOR J = Q2 + 1 TO Q1: C(J) = 0: S(J) = 0: NEXT J
```

[Note: we keep showing these detailed, low level, stage by stage listings of 8 and 16 point FFTs because: it's one thing to explain how the partial transforms increase in size and decrease in number, twiddle factors change with each stage, etc., etc., and quite another to handle all these quasi-constant variables when writing code. When we lay-out all the *nested* loops sequentially as in the above listing it becomes apparent exactly how they change from stage to stage as well as between partial transforms—they "jump-out at you." Now, this book is about FFT applications; but, the FFT is just a computer algorithm. All FFT applications are applications of an algorithm.... Listen carefully—it's apparent there are many ways to write FFT software, and if you continue working with the FFT you will eventually come to see that, for many applications, it's best to write your own software (this will be amply illustrated in the following chapters). When the day comes that you tackle your first FFT, you will be well advised to lay-out all the *loops* as we have done above. Let's look at the example before us:

From Figs. 7.3, 7.5 and 7.6; and the partial routines presented so far, it's apparent that in each succeeding stage the J counter will count from 0 to 1; 0 to 2; 0 to 4; etc. (remember we have already done the zero stage). With the number of data points $Q = 2^N$ we know there will be N stages of computation; so, if we let M count 1 to N, it will count the stages of computation. Furthermore, $QP = 2^M$ will equal the size of each partial transform in each stage, and $QPI = 2^{(N-M)}$ will equal the number of partial transforms at that stage. If we review the 16 point transform on the preceding page (with all its individual loops) we will find that these two variables are most of what we need to produce a *complete* PFFFT. Having made these observations it will be apparent that the new algorithm we are looking for can be written as follows:

```
10 ' ***  FFT07-01 *** POSITIVE FREQUENCY FFT ***
12 CLS : PRINT "INPUT NUMBER OF DATA POINTS AS 2^N"
14 INPUT "N = "; N
16 Q = 2^N: N1 = N-1: Q1 = Q-1: Q2 = Q/2: Q3 = Q2-1: Q4 = Q/4: Q5 = Q4-1
18 Q8 = Q/8: Q9 = Q8 - 1: Q16 = Q/16
20 DIM Y(Q), C(Q), S(Q), KC(Q2), KS(Q2)
30 PI = 3.14159265358979#: P2 = PI * 2: K1 = P2/Q

39 ' ****  MAIN MENU  ****
40 CLS : PRINT SPC(30); "MAIN MENU": PRINT : PRINT
60 PRINT SPC(5); "1 = ANALYZE Q/2 COMPONENT FUNCTION": PRINT
62 PRINT SPC(5); "2 = EXIT": PRINT
70 PRINT SPC(10); "MAKE SELECTION: ";
80 A$ = INKEY$: IF A$ = "" THEN 80
90 A = VAL(A$): ON A GOSUB 600, 900
92 GOTO 40

100 '               *** TRANSFORM ***
108 ' *** TRANSFORM STAGE 0  ***
110 C(0) = (S(0) + S(Q2))/2: C(1) = (S(0) - S(Q2))/2
112 FOR I = 1 TO Q3: I2 = 2*I: INDX = 0
114 FOR J = 0 TO N1: IF I AND 2^J THEN INDX = INDX + 2^(N-2-J)
```

The PFFFT 81

```
116 NEXT J
118 C(I2) = (S(INDX) + S(INDX+Q2))/2: C(I2+1) = (S(INDX) - S(INDX+Q2))/2
120 NEXT I
122 FOR I = 0 TO Q1: S(I) = 0: NEXT I ' CLEAR SINE COMPONENTS
'  *********  SUM REMAINING STAGES  ************
124 FOR M = 1 TO N1: QP = 2^M: QPI = 2^(N1-M): K0 = QPI*K1
126   FOR K = 0 TO QPI - 1
128     FOR J = 0 TO QP/2: J0 = J + (2*K*QP)
130       J1 = J0 + QP: JI = J1 - (2*J): K2 = K0*J: K3 = K0*JI
132       CTEMP1 = C(J0) + C(J1)*COS(K2) - S(J1)*SIN(K2)
134       STEMP1 = S(J0) + C(J1)*SIN(K2) + S(J1)*COS(K2)
136       CTEMP2 = C(J0) + C(J1)*COS(K3) + S(J1)*SIN(K3)
138       S(JI) = (C(J1)*SIN(K3) - S(J1)*COS(K3) - S(J0))/2
140       C(J0) = CTEMP1 /2: S(J0) = STEMP1 /2: C(JI) = CTEMP2 /2
142     NEXT J
144   NEXT K
146 NEXT M
148 FOR J = Q2+1 TO Q1: C(J) = 0: S(J) = 0: NEXT J 'CLEAR UPPER ARRAY
150 SK1 = 2 ' SET PRINT SCALE FACTOR

'  *******  PRINT OUTPUT  *******
200 FOR Z = 0 TO Q3' PRINT OUTPUT
210 PRINT USING "###"; Z; : PRINT "   ";
212 PRINT USING "+##.#####"; SK1 * C(Z); : PRINT "   ";
214 PRINT USING "+##.#####"; SK1 * S(Z); : PRINT "    ";
216 PRINT USING "###"; Z + Q2; : PRINT "   ";
218 PRINT USING "+##.#####"; SK1 * C(Z + Q2); : PRINT "   ";
220 PRINT USING "+##.#####"; SK1 * S(Z + Q2)
230 NEXT Z
240 PRINT : INPUT "ENTER TO CONTINUE"; A$
250 RETURN

600 ' ***  Q/2 COMPONENT FUNCTION  ***
602 CLS : PRINT : PRINT
604 PRINT SPC(10); "1 = TRIANGLE WAVE": PRINT SPC(10); "2 = NYQUIST"
605 A$ = INKEY$: IF A$ = "" THEN 605
606 A = VAL(A$): IF A <> 1 AND A <> 2 THEN 602
608 PRINT "PREPARING DATA - PLEASE WAIT"
610 ON A GOSUB 650, 660
612 SK1 = 1: GOSUB 200
614 PRINT : INPUT "INPUT DATA READY - ENTER TO CONTINUE"; A$
620 GOSUB 100 'TAKE TRANSFORM
630 RETURN 'BACK TO MAIN MENU
650 ' ***  GENERATE Q/2 COMPONENT TRIANGLE  ***
652 FOR I = 0 TO Q1: Y(I) = 0: C(I) = 0: S(I) = 0 ' CLEAR DATA POINTS
654 FOR J = 1 TO Q2 STEP 2: S(I) = S(I) + COS(K1 * J * I) / (J * J): NEXT
656 NEXT
658 RETURN ' BACK TO GENERATE FUNCTION ROUTINE
660 D = 1' ***  GENERATE NYQUIST COMPONENT  ***
662 FOR I = 0 TO Q1: Y(I) = 0: C(I) = 0: S(I) = D ' CLEAR & DATA POINT
666 D = D * (-1): NEXT ' NEGATE DATA POINT & REPEAT
668 RETURN ' BACK TO GENERATE FUNCTION ROUTINE

'  **********
900 STOP
```

We're obviously generating twiddle factors *on-the-fly* here so you might want to look into optimizing this routine. This will be our algorithm of choice for transforming real valued time domain data; for, time wasted transforming and manipulating negative frequencies is eliminated—this is precisely what we need for applications such as *spectrum analysis*. After working with this algorithm for a while you may find it difficult to go back to the *conventional* FFT.... The biggest drawback at the moment is that this routine works only in the *forward* direction.

CHAPTER 8

THE INVERSE PFFFT

8.0 INTRODUCTION

In the inverse *DFT* we simply generate sinusoids (with their amplitudes prescribed by the phasor components of the DFT) and sum together the contributions of each of these sinusoids at every data point in the time domain. This is very intuitive—we *understand* what's going on. When we perform the inverse FFT we accomplish the same result as the inverse DFT, but the route by which we get there is considerably less intuitive. In *Understanding the FFT* we didn't actually derive the inverse FFT—there was no need to do so and precious little interest in such a derivation. We had already shown that the inverse FFT was the same algorithm (essentially) as the forward FFT.

For the inverse PFFFT, however, it's *obvious* we can't use the same algorithm to transform from frequency to time domain—a very cursory consideration of the preceding algorithm reveals the PFFFT starts with Q time domain data points and produces Q/2 + 1 frequency domain components. If we attempted to *inverse* transform data via this algorithm we would wind up with (Q/4 + 1) time domain data points, and that obviously doesn't work. Even if we synthesized the negative frequencies before entering this routine we would only obtain (Q/2 + 1) data points...we really *can't* get there from here.

I'm afraid there's no alternative to dragging you through another delightful experience—sorry about that ☺ *NOTE: Apparently, there are a multitude of ways to interpret and construct FFTs (and inverse FFTs); however, just because there are a lot of different ways to do it right doesn't mean it's easy—there are even* more *ways to do it wrong!*

8.1 THE INVERSE PFFFT

To construct an inverse PFFFT we will work the forward PFFFT in *reverse*. The basic idea to this approach is as follows: start with the final transform stage and work backward to the preceding stage (where we summed two partial transforms to get the final results—see Fig. 8.1).

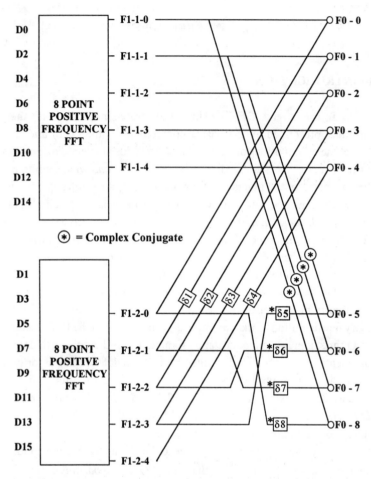

Figure 8.1 - 16 Data Point PFFFT Butterfly (Final Stage)

Since we create two components in the output by summing pairs of components from the preceding stages (F_{0-0} and F_{0-8}, for example, are both formed by summing F_{1-1-0} and F_{1-2-0}—see above); then, given the output components, we should be able to solve the equations used to obtain these results for the components of the preceding stage. This will become clear in the following paragraphs, and it will also be clear we may continue backing through the PFFFT in this manner, eventually ending up with exactly Q *time domain* data points. Okay, how are we going to do this?

Figure 8.1 shows the final stage of a 16 data point PFFFT. The

The Inverse PFFFT

first three stages of computation have resulted in two blocks shown as 8-point partial transforms. This diagram shows how the two partial transforms are combined to form the nine frequencies of a 16 data point PFFFT. The final frequency components are formed as follows:

and:
$$F_{0-0} = F_{1-1-0} + F_{1-2-0} \tag{8.1}$$
$$F_{0-8} = F_{1-1-0}{}^* + [\delta 8]\, F_{1-2-0}{}^* \tag{8.2}$$

where the * indicates complex conjugates and $[\delta 8]$ indicates the usual phase rotation necessary for the time-shifted transform.[1] Similarly, the other components are formed as follows:

$$F_{0-1} = F_{1-1-1} + [\delta 1]F_{1-2-1} \tag{8.3}$$
$$F_{0-7} = F_{1-1-1}{}^* + [\delta 7]F_{1-2-1}{}^* \tag{8.4}$$

$$F_{0-2} = F_{1-1-2} + [\delta 2]F_{1-2-2} \tag{8.5}$$
$$F_{0-6} = F_{1-1-2}{}^* + [\delta 6]F_{1-2-2}{}^* \tag{8.6}$$

$$F_{0-3} = F_{1-1-3} + [\delta 3]F_{1-2-3} \tag{8.7}$$
$$F_{0-5} = F_{1-1-3}{}^* + [\delta 5]F_{1-2-3}{}^* \tag{8.8}$$

and:
$$F_{0-4} = F_{1-1-4} + [\delta 4]F_{1-2-4} \tag{8.9}$$

Our objective is to work backwards from the F_{0-X} components to the two sets of F_{1-X-X} components and to do this we will treat each pair above as simultaneous equations. Let's arbitrarily use (8.3) and (8.4) as examples, recognizing all pairs will behave the same way:

$$\{F_{0-1}\} - \{F_{0-7}\} = F_{1-1-1} - F_{1-1-1}{}^* + [\delta 1]F_{1-2-1} - [\delta 7]F_{1-2-1}{}^* \tag{8.10}$$

Let's ignore the F_{1-2-1} components for the moment and concentrate on the F_{1-1-1} components. We know the only difference between these two terms is the negated imaginary component, so subtracting $F_{1-1-1}{}^*$ from F_{1-1-1} removes the *real* component and yields twice the *imaginary* component. Now, when we work "inside" the FFT, we handle the real and imaginary components separately, so let's look at the equations for the real parts (i.e., the cosine components) of F_{0-1} and F_{0-7}:

$$\mathrm{Re}\{F_{0-1}\} = (\mathrm{Re}\{F_{1-1-1}\} + \mathrm{Cos}\beta_1 \mathrm{Re}\{F_{1-2-1}\} - \mathrm{Sin}\beta_1 \mathrm{Im}\{F_{1-2-1}\})/2 \tag{8.11}$$
$$\mathrm{Re}\{F_{0-7}\} = (\mathrm{Re}\{F_{1-1-1}{}^*\} + \mathrm{Cos}\beta_7 \mathrm{Re}\{F_{1-2-1}{}^*\} - \mathrm{Sin}\beta_7 \mathrm{Im}\{F_{1-2-1}{}^*\})/2 \tag{8.12}$$

---

[1] We ignore the requirement of dividing the summed components by 2 in the forward transform—following this derivation will be difficult enough!

When we subtract (8.12) from (8.11), F_{1-1-1} disappears completely leaving:

$$\text{Re}\{F_{0-1}\} - \text{Re}\{F_{0-7}\} = (\text{Re}\{[\delta 1]F_{1-2-1} - [\delta 7]F_{1-2-1}^*\})/2 \tag{8.13}$$

Very well then, we have eliminated the F_{1-1-1} term from these equations—we now need to look at what's left [i.e., the F_{1-2-1} terms]:

$$\text{Re}\{[\delta 1]F_{1-2-1} - [\delta 7]F_{1-2-1}^*\} = (\text{Cos}\beta_1 \text{Re}\{F_{1-2-1}\} - \text{Sin}\beta_1 \text{Im}\{F_{1-2-1}\} - \text{Cos}\beta_7 \text{Re}\{F_{1-2-1}^*\} + \text{Sin}\beta_7 \text{Im}\{F_{1-2-1}^*\})/2 \tag{8.14}$$

β_1 and β_7 are the *twiddle factor* angles, and because of the way we generate these angles, β_7 is equal to $\pi - \beta_1$ (i.e., they are generated by dividing π into N equal increments). Furthermore, $\text{Cos}(\pi - \beta_1) = -\text{Cos}(\beta_1)$ and $\text{Sin}(\pi - \beta_1) = \text{Sin}(\beta_1)$. Substituting these Eqn.(8.14) becomes:

$$\begin{aligned}\text{Re}\{[\delta 1]F_{1-2-1} - [\delta 7]F_{1-2-1}^*\} &= [\text{Re}\{F_{1-2-1}\}(\text{Cos}\beta_1 + \text{Cos}\beta_1) - \text{Im}\{F_{1-2-1}\}(\text{Sin}\beta_1 + \text{Sin}\beta_1)]/2 \\ &= [\text{Re}\{F_{1-2-1}\}(2\text{Cos}\beta_1) - \text{Im}\{F_{1-2-1}\}(2\text{Sin}\beta_1)]/2 \\ &= [\text{Re}\{F_{1-2-1}\}(\text{Cos}\beta_1) - \text{Im}\{F_{1-2-1}\}(\text{Sin}\beta_1)] \end{aligned} \tag{8.15}$$

[Note: (8.15) is just the component of phasor F_{1-2-1} that's projected onto the real axis when it is rotated by an angle of β_1.]

Okay, let's now look at the imaginary terms for F_{0-1} and F_{0-7}:

$$\text{Im}\{F_{0-1}\} = (\text{Im}\{F_{1-1-1}\} + \text{Sin}\beta_1 \text{Re}\{F_{1-2-1}\} + \text{Cos}\beta_1 \text{Im}\{F_{1-2-1}\})/2 \tag{8.16}$$

$$\text{Im}\{F_{0-7}\} = (\text{Im}\{F_{1-1-1}^*\} + \text{Sin}\beta_7 \text{Re}\{F_{1-2-1}^*\} + \text{Cos}\beta_7 \text{Im}\{F_{1-2-1}^*\})/2 \tag{8.17}$$

We know the imaginary parts of F_{1-1-1}^* and F_{1-2-1}^* are just the negated imaginary parts of F_{1-1-1} and F_{1-2-1}. Therefore, (8.17) becomes:

$$\text{Im}\{F_{0-7}\} = (-\text{Im}\{F_{1-1-1}\} + \text{Sin}\beta_7 \text{Re}\{F_{1-2-1}\} - \text{Cos}\beta_7 \text{Im}\{F_{1-2-1}\})/2 \tag{8.18}$$

To remove the F_{1-1-1} term from these equations it's apparent that we must *add* (8.16) and (8.18):

$$\text{Im}\{F_{0-1}\} + \text{Im}\{F_{0-7}\} = (\text{Sin}\beta_1 \text{Re}\{F_{1-2-1}\} + \text{Sin}\beta_7 \text{Re}\{F_{1-2-1}\} + \text{Cos}\beta_1 \text{Im}\{F_{1-2-1}\} - \text{Cos}\beta_7 \text{Im}\{F_{1-2-1}\})/2 \tag{8.19}$$

And as before, $\text{Cos}\beta_7 = -\text{Cos}\beta_1$ and $\text{Sin}\beta_7 = \text{Sin}\beta_1$:

$$\begin{aligned}\text{Im}\{F_{0-1}\} + \text{Im}\{F_{0-7}\} &= [\text{Re}\{F_{1-2-1}\}(2\text{Sin}\beta_1) + \text{Im}\{F_{1-2-1}\}(2\text{Cos}\beta_1)]/2 \\ &= \text{Re}\{F_{1-2-1}\}(\text{Sin}\beta_1) + \text{Im}\{F_{1-2-1}\}(\text{Cos}\beta_1) \end{aligned} \tag{8.20}$$

This, of course, is the component of phasor F_{1-2-1} that's projected onto the

The Inverse PFFFT

imaginary axis when it's rotated by an angle of β_1. Together, (8.15) and (8.20) are the rotated phasor F_{1-2-1}. If we rotate the phasor represented by (8.15) and (8.20) backward by $-\beta_1$ we will have the thing we were looking for—the un-rotated phasor F_{1-2-1}. The real part of F_{1-2-1} is therefore:

$$Re\{F_{1-2-1}\} = [Re\{F_{0-1}\} - Re\{F_{0-7}\}]Cos\beta_1 + [Im\{F_{0-1}\}+Im\{F_{0-7}\}]Sin\beta_1 \quad (8.21)$$

and the imaginary part is:

$$Im\{F_{1-2-1}\} = [Im\{F_{0-1}\} + Im\{F_{0-7}\}]Cos\beta_1 - [Re\{F_{0-1}\}-Re\{F_{0-7}\}]Sin\beta_1 \quad (8.22)$$

So then, the forward transform is given by Eqn.(8.3):

$$F_{0-1} = [F_{1-1-1} + [\delta 1]F_{1-2-1}]/2 \quad (8.3)$$
$$= [F_{1-1-1} + e^{i\beta 1}F_{1-2-1}]/2 \quad (8.3A)$$

while, from (8.21) and (8.22) the inverse transform for F_{1-2-1} is given by:

$$F_{1-2-1} = [F_{0-1} - F_{0-7}{}^*]e^{-i\beta 1} \quad (8.23)$$

These are similar equations but, once we know the coefficients for F_{1-2-1} the equation for F_{1-1-1} is even closer:

$$F_{1-1-1} = 2F_{0-1} - e^{i\beta 1}F_{1-2-1} \quad (8.24)$$

which is derived directly from (8.3A) of course. From Figure 8.1 and Eqns. (8.1) through (8.9) it's apparent that *all* the harmonic components of the F_{1-x-x} stages may be determined from similar equations (allowing a $[\delta 0]$ phase term in Eqn. (8.1) equal to zero). Only Equation (8.9) is different, but it will be solved by the same set of equations, since we recognize that it is generated by either equation of the preceding pairs. If we allow it to be used twice in the inverse transform routine it will produce the correct phasor for F_{1-2-4} which may be used to find F_{1-1-4}. From the above we may write our inverse transform routine via the following rules:

1. Subtract the real (i.e., cosine) term of $F_{0-(8-x)}$ from the real term of F_{0-x}: $M_{CT} = Re\{F_{0-X}\} - Re\{F_{0-(8-X)}\}$.
2. Add the imaginary (i.e., sine) part of $F_{0-(8-x)}$ to the imaginary part of F_{0-x}: $M_{ST} = Im\{F_{0-X}\} + Im\{F_{0-(8-X)}\}$.
3. The cosine term for F_{1-2-X} then equals $M_{CT}Cos\beta_X + M_{ST}Sin\beta_X$.
4. The sine term is equal to $M_{ST}Cos\beta_X - M_{CT}Sin\beta_X$.

Note: Steps 3. and 4. are simply the operation of rotating the phasor backward by β_X. $F_{1\text{-}1\text{-}X}$ is then obtained by subtracting the rotated phasor $F_{1\text{-}2\text{-}X}$ from the phasor $F_{0\text{-}X}$ as follows:

5. The cosine term is given by: $\text{Re}\{F_{0\text{-}X}\} - \text{Re}\{F_{1\text{-}2\text{-}X}\}\cos\beta_X + \text{Im}\{F_{1\text{-}2\text{-}X}\}\sin\beta_X$.
6. The sine term is: $\text{Im}\{F_{0\text{-}X}\} - \text{Re}\{F_{1\text{-}2\text{-}X}\}\sin\beta_X - \text{Im}\{F_{1\text{-}2\text{-}X}\}\cos\beta_X$.

That's pretty much all there is to it; however, in the previous chapter we talked about writing FFT routines, constructing loops for each stage of computation, and cascading these stages into a final, complete, FFT routine...and how much fun that exercise would be. We have before us a golden opportunity to test that proposition by writing our own inverse PFFFT...to see if all the tedium is necessary or not. It might help to note that, so far as array data management is concerned, the upper half of each partial transform will always be empty (except for the Nyquist term).

In the following program a version of the inverse PFFFT is created from the rules given above, but you should not expect to wind up with identical code:

```
8   '***********************************************
10  ' ***   FFT08-01  ***  POSITIVE FREQUENCY FFT  ***
11  '***********************************************
12  CLS : PRINT "INPUT NUMBER OF DATA POINTS AS 2^N"
14  INPUT "N = "; N
16  Q = 2^N: N1 = N - 1: Q1 = Q - 1: Q2 = Q/2: Q3 = Q2 - 1: Q4 = Q/4
18  Q5 = Q4 - 1: Q8 = Q/8: Q9 = Q8 - 1: Q34 = Q2 + Q4: Q16 = Q/16
20  DIM Y(Q), C(Q), S(Q), KC(Q2), KS(Q2)
30  PI = 3.14159265358979#: P2 = PI*2: K1 = P2/Q
32  FOR I = 0 TO Q3: KC(I) = COS(I*K1): KS(I) = SIN(I*K1): NEXT I

    '********************************
    '********   MAIN MENU   ********
    '********************************
40  CLS : PRINT SPC(30); "MAIN MENU": PRINT : PRINT
60  PRINT SPC(5); "1 = ANALYZE Q/2 COMPONENT FUNCTION": PRINT
62  PRINT SPC(5); "2 = INVERSE TRANSFORM": PRINT
64  PRINT SPC(5); "3 = PRINT RESULTS": PRINT
66  PRINT SPC(5); "4 = EXIT": PRINT
70  PRINT SPC(10); "MAKE SELECTION: ";
80  A$ = INKEY$: IF A$ = "" THEN 80
82  PRINT A$
90  A = VAL(A$): ON A GOSUB 600, 200, 350, 900
92  GOTO 40
```

The Inverse PFFFT

```
98  '           ******************************
100 '           ***    FORWARD TRANSFORM   ***
102 '           ******************************
106 '               ***  TRANSFORM STAGE 1  ***
108 T9 = TIMER
110 C(0) = (S(0) + S(Q2))/2: C(1) = (S(0) - S(Q2))/2
112 FOR I = 1 TO Q3: I2 = 2*I: INDX = 0
114 FOR J = 0 TO N1: IF I AND 2^J THEN INDX = INDX + 2^(N - 2 - J)
116 NEXT J
118 C(I2) = (S(INDX)+S(INDX+Q2))/2: C(I2+1) = (S(INDX) - S(INDX+Q2))/2
120 NEXT I
122 FOR I = 0 TO Q1: S(I) = 0: NEXT I'  ZERO SINE TERMS
  '        *********  REMAINING STAGES  **********
124 FOR M = 1 TO N1: QP = 2^M: QPI = 2^(N1 - M)
126   FOR K = 0 TO QPI - 1
128     FOR J = 0 TO QP/2: J0 = J + (2*K*QP): J1 = J0 + QP: K2 = QPI*J
130       JI = J1 - (2*J)
132       CTEMP1 = C(J0) + C(J1)*KC(K2) - S(J1)*KS(K2)
134       STEMP1 = S(J0) + C(J1)*KS(K2) + S(J1)*KC(K2)
136       CTEMP2 = C(J0) - C(J1)*KC(K2) + S(J1)*KS(K2)
138       S(JI) = (C(J1)*KS(K2) + S(J1)*KC(K2) - S(J0))/2
140       C(J0) = CTEMP1/2: S(J0) = STEMP1/2: C(JI) = CTEMP2/2
142     NEXT J
144   NEXT K
146 NEXT M
148 FOR J = Q2 + 1 TO Q1: C(J) = 0: S(J) = 0: NEXT J
150 T9 = TIMER - T9: SK1 = 2
152 RETURN

REM ******************************************************
REM *              INVERSE TRANSFORM                     *
REM ******************************************************
200 PRINT : T9 = TIMER: SK1 = 1
202 FOR M = N1 TO 1 STEP -1'  LOOP FOR STAGES OF COMPUTATION
204   QP2 = 2^(M): QP = INT(QP2/2): QP4 = 2*QP2: QPI = 2^(N1 - M)
206   FOR I = 0 TO Q - (QP2) STEP QP4
208     FOR J = 0 TO QP: KI = J + I: KT = J*QPI: KJ = QP2 + KI
212       MCT = C(J + I) - C(I + QP2 - J): MST = S(J + I) + S(I + QP2 - J)
214       CTEMP = MCT * KC(KT) + MST * KS(KT)
216       STEMP = MST * KC(KT) - MCT * KS(KT)
218       CTEMP2 = (2*C(J + I)) - CTEMP*KC(KT) + STEMP*KS(KT)
220       S(KI) = (2*S(J + I)) - CTEMP*KS(KT) - STEMP*KC(KT)
222       C(KJ) = CTEMP: S(KJ) = STEMP: C(KI) = CTEMP2
224     NEXT J
226   NEXT I
228 NEXT M
229 '     ********  FINAL STAGE  ********
230 FOR I = 0 TO Q3: I2 = 2*I: INDX = 0
232 FOR J = 0 TO N1: IF I AND 2^J THEN INDX = INDX + 2^(N - 2 - J)
234 NEXT J
236 S(INDX) = C(I2) + C(I2 + 1): S(INDX + Q2) = C(I2) - C(I2 + 1)
238 NEXT I
240 FOR I = 0 TO Q1: C(I) = 0: NEXT I
242 T9 = TIMER - T9
244 RETURN
```

```
'        *********************************
'        *******   PRINT OUTPUT   *******
'        *********************************
350 CLS : PRINT SPC(16); "*****    "; NM$; "  *****"
352 PRINT "   FREQ       F(COS)       F(SIN)           FREQ       F(COS)       F(SIN)"
354 PRINT
362 TZT = 20: FOR Z = 0 TO Q4' PRINT OUTPUT
364 PRINT USING "####"; Z; : PRINT "   ";
366 PRINT USING "+##.#####"; SK1 * C(Z); : PRINT "   ";
368 PRINT USING "+##.#####"; SK1 * S(Z); : PRINT "      ";
370 PRINT USING "######"; Z + Q4; : PRINT "   ";
372 PRINT USING "+##.#####"; SK1 * C(Z + Q4); : PRINT "   ";
374 PRINT USING "+##.#####"; SK1 * S(Z + Q4)
376 IF Z > TZT THEN TZT = TZT + 20: INPUT A$
378 NEXT Z
380 PRINT : PRINT "T = "; T9: INPUT "ENTER TO CONTINUE"; A$
382 RETURN

600 ' ***  GENERATE SINE WAVE COMPONENT   ***
602 CLS : PRINT : PRINT
604 INPUT "PLEASE SPECIFY FREQUENCY"; F9
608 WTMSG$ = "                        PREPARING DATA - PLEASE WAIT"
610 FOR I = 0 TO Q1: C(I) = 0: S(I) = SIN(F9 * K1 * I): NEXT I
618 T9 = TIMER
620 GOSUB 100 'TAKE TRANSFORM
622 GOSUB 350 ' DISPLAY TRANSFORM
630 RETURN 'BACK TO MAIN MENU
' **********
900 STOP ' THAT'S ALL FOLKS
```

Note that, again, we have combined the bit-reversal routine with the final rotate and sum stage of the transform. When was the last time you had so much fun so cheaply?

CHAPTER 9

THE SPECTRUM ANALYZER

9.0 INTRODUCTION

It might seem that spectrum analysis would be the simplest and most direct application of the FFT for, after all, that is what an FFT does—it finds the spectrum of a time domain signal. Unfortunately, as we shall see in this chapter, practical spectrum analysis turns out to be neither simple nor direct. We choose the spectrum analyzer as an application not because it is simple, but because it is enlightening. We will emerge from this chapter with a much better understanding of Fourier transforms.

9.1 THE NATURE OF THE BEAST

Spectrum analysis brings us into conflict with the real world, for here we don't have the luxury of analyzing mathematical functions that are made-to-order (so to speak). In the practical realm of spectrum analysis, the signals are generated by and for completely independent criteria, and the customer who purchases an instrument to analyze his spectrum expects it to accurately measure and display the spectrum of his *signal*—not artifacts of the DFT process.

We begin our investigation using the program we developed in the last chapter (i.e., FFT08.01). We will study a simple input—a single sinusoid (a sine wave) with its frequency selected by the user. This single sinusoid input is ideal for our task—it produces a single point in the frequency domain so it will be easy to understand what we see. So, what could possibly go wrong?

At this early stage in our discussion we need waste no time on long-winded theoretical discussions—as they say in show-biz, let's *cut directly to the chase*[1].

---

[1] Hollywood discovered long ago that, after struggling interminably with no story at all, a 10 minute "chase scene" guaranteed a box office success. Shortly thereafter everyone figured out what they were doing—hence the expression. These expressions can't stand up to the overuse they're given but, used sparingly, they're effective.

Run FFT08.01 (from the previous chapter) and select N = 6 for 64 data points (this will make it easy to see and understand the data printout). Next select a frequency of 16 (i.e., half way to the Nyquist). The computer will churn through the analysis and a truly profound thing will happen—a single data point will show up on the printout at $f = 16$ (if you find this result surprising, a little review work might be in order :-).

FREQ	F(COS)	F(SIN)	FREQ	F(COS)	F(SIN)
0	-0.0000	+0.0000	16	+0.0000	+1.0000
1	-0.0000	+0.0000	17	-0.0000	-0.0000
2	-0.0000	+0.0000	18	-0.0000	-0.0000
3	-0.0000	+0.0000	19	-0.0000	-0.0000
4	-0.0000	+0.0000	20	-0.0000	-0.0000
5	-0.0000	+0.0000	21	-0.0000	-0.0000
6	-0.0000	+0.0000	22	-0.0000	-0.0000
7	-0.0000	+0.0000	23	-0.0000	-0.0000
8	-0.0000	+0.0000	24	-0.0000	-0.0000
9	-0.0000	+0.0000	25	-0.0000	-0.0000
10	-0.0000	+0.0000	26	-0.0000	-0.0000
11	-0.0000	+0.0000	27	-0.0000	-0.0000
12	-0.0000	+0.0000	28	-0.0000	-0.0000
13	-0.0000	+0.0000	29	-0.0000	-0.0000
14	-0.0000	+0.0000	30	-0.0000	-0.0000
15	-0.0000	+0.0000	31	-0.0000	-0.0000
16	+0.0000	+1.0000	32	-0.0000	-0.0000

TIME = 5.859375E-02

Okay, we know the DFT (and FFT) can only analyze waveforms into orthogonal components. That is, frequencies that fit perfectly into the time domain interval; however, we also know that, in the real world, the sinusoids we input to our spectrum analyzer may be any frequency whatsoever! So, what happens when we analyze some frequency that's not a perfect match for our computer algorithm? What happens when we analyze a sinusoid that doesn't match-up with the harmonics used in the analysis?

Run FFT08.01 again and this time select a frequency of *16.1*. Pretty messy, hey? Messy, but nonetheless, the basic sine wave component is still there...we knew *something* would have to "give." Look at it this way: it would seem not unlikely that the amplitude of a sinusoid (whose frequency was between the frequencies of our analytical scheme) would be sort of averaged between the frequencies on either side. Okay, but what's that *garbage* all over the spectrum?

Run the analysis yet again, and select a frequency of *16.2*. Ummm...things are *not* getting better. We can still tell there's a signal in there somewhere, but the garbage is getting worse.... Let's cut directly to the chase here. Run the analysis with a frequency midway between the two analyzing components—run it at *16.5* Hz. The result of this analysis is shown in the printout below, and it's not good!

The Spectrum Analyzer

FREQ	F(COS)	F(SIN)	FREQ	F(COS)	F(SIN)
0	+0.0298	+0.0000	16	+0.6361	-0.0000
1	+0.0299	+0.0000	17	-0.6376	+0.0000
2	+0.0303	-0.0000	18	-0.2137	+0.0000
3	+0.0310	-0.0000	19	-0.1294	+0.0000
4	+0.0321	-0.0000	20	-0.0936	+0.0000
5	+0.0335	-0.0000	21	-0.0739	+0.0000
6	+0.0355	-0.0000	22	-0.0616	+0.0000
7	+0.0380	-0.0000	23	-0.0533	+0.0000
8	+0.0413	-0.0000	24	-0.0474	+0.0000
9	+0.0457	-0.0000	25	-0.0431	+0.0000
10	+0.0516	-0.0000	26	-0.0399	+0.0000
11	+0.0600	-0.0000	27	-0.0375	+0.0000
12	+0.0723	-0.0000	28	-0.0357	+0.0000
13	+0.0920	-0.0000	29	-0.0344	+0.0000
14	+0.1278	-0.0000	30	-0.0335	+0.0000
15	+0.2122	-0.0000	31	-0.0330	+0.0000
16	+0.6361	-0.0000	32	-0.0328	-0.0000

TIME = 0.0000

We are generating a *sine* wave to test our FFT analyzer, but the *sine* component has disappeared completely. There's nothing but cosine garbage spattered *all over* the spectrum. We're supposed to sell this thing to people who want to see the sinusoidal components in their signal, but it has completely lost our input component. I wouldn't buy this thing myself! Ahhhhhhh...Mission Control, we *may* have a problem.

Why does the transform of a single sinusoid come out this way? Well, that *is* the question; but, before we rush off on that tangent let's complete the above illustration. If you *reconstruct* the 16.5 Hz signal from the garbage shown above, *you will get a faithful reproduction of the original input!* Even though no *sine* component appears at 16.5 Hz (there *is* no 16.5 Hz component), the cosine component *garbage* yields a sine wave on reconstruction. Apparently this is a legitimate transform(?), but it's like our FFT has "busted a spring" trying to come up with a transform for a frequency it doesn't have! That kind of commitment is admirable, but this *busted spring transform* is useless as a spectrum analyzer. Ahhhhh...Mission Control, we *do* have a problem.

Now let's look at why. First of all, the size of the data array (e.g., 64 data points, 128 data points, etc.,) isn't completely arbitrary. Sampling rate must be twice the highest frequency in the signal (we will talk about this in detail later) and we must digitize long enough to include the lowest frequency of interest. Other things will influence the decision (e.g., the use of a round number 2^N data points); still, the lowest and highest frequencies of concern will determine the *minimum* data array size (i.e., sampling rate multiplied by data interval—see Fig. 9.1). Now, from the orthogonality requirement, we must analyze this data using harmonics which *fit* into the data interval (i.e., we must use integer multiples of the *fundamental* sinusoid that fit exactly into the domain of the digitized function—Fig. 9.1); however, it's clear the problem with the N+1/2 Hz frequency components *dropping out* is related to the use of these integer multiple frequency components...hmmm.

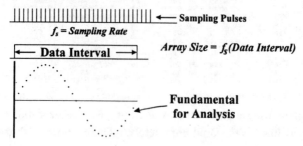

Figure 9.1 - Data Array Size

9.2 FRACTIONAL FREQUENCY ANALYSIS

It's pretty obvious we will have to break out of these restrictions if we hope to build a better mousetrap...but how? Well, I've always wondered if we're *really* forced to use this integer multiple frequency configuration? [I can just hear the folks back in Peoria right now... *"What?!! What are you talking about Andy?!! Have you lost your mind?! We've gone through one and a half books constantly hammering at the necessity of using orthogonal components!"*]

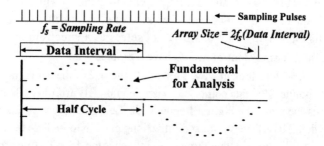

Figure 9.2 - Half Hertz Analysis

In Figure 9.2 we place the data shown in Fig. 9.1 into an array of twice the length (i.e., if our array in Fig. 9.1 was 64 data points, we would use those data points for the first half of a 128 data point array and *pack*

The Spectrum Analyzer

the second half with zeros). We know we can take the FFT of this 128 point array just as we did the 64 point array, and we know the inverse transform will reproduce the data we started with (i.e., 64 data points followed by 64 zeros). Let's ignore the 64 zeros for now and look at what has happened to the relationship of the fundamental (and harmonics) of the analytical configuration. Look at Fig 9.2 again—as far as the 64 data points are concerned we are now analyzing the data every 1/2 Hz!

We can see where this is going; but, is *this* transform the same as the one we obtained before? Obviously it isn't—we have frequency components we didn't have before, but is this a "correct" spectrum for our 64 data points? There can be no question this is a legitimate transform... but is it the true spectrum of the input signal? Will it accurately display sinusoid components that really exist in the input signal (and not display components that *don't* exist in the input signal)? A more immediate question is whether this 1/2 Hz transform will actually recover the *fractional frequency* we lost in the conventional transform?

Let's cut to the chase scene here. What we really want to know is this: "Can we develop a 1/2 Hz FFT that will cause our spectrum analyzer to detect *half-frequency* components?" Well, we can certainly make an analyzer that will *look for* 1/2 Hz components! All we need to do is increase the array size to the next *power of 2* and fill the second half of the array with zeros. [There is, of course, another way, but let's keep it simple—at least until we can figure out if this is going to work.] The real question is whether this will accurately display 1/2 Hz components.

Okay, let's cut to the chase here—just run program FFT9.02 (Appendix 9.4). This program starts-up with a menu of *fractional-frequency modes*. Any selection jumps to a spectrum analyzer display (with a single frequency input signal). We can increment or decrement the frequency by 0.1 Hz increments using the cursor control (arrow) keys. Okay, let's find out what we really want to know—select the second illustration which implements *1/2 Hz analysis* of the data. We note that now, with this 1/2 Hz analysis, even with a nice round integer frequency the *garbage* is present...but let's not worry about that now.... Using the cursor key, step the frequency up 0.5 Hz (i.e., press the *right* cursor arrow five times). Watch the spectrum change with each step. Even though the amplitude still changes with each step, it only varies by a few percent—it certainly never vanishes. Okay, we just might be moving forward here....

Let's pause here to talk about this *extended array* data and its transform. Fig. 9.2 (and a little mental juggling) reveals that the *even* components of the 128 point analysis are *identical* to the components of the original 64 point analysis (recognizing an amplitude *scale factor* change of course). So, if the *even* frequency components are identical to the original analysis, what are the *odd* frequency components doing? If the even frequency components represent the original spectrum, how much do these additional *odd* components distort the spectrum?

[Consider this: if we input an integer frequency sinusoid, the even components will be identical to the original analysis, but on reconstruction the transform must know to stop generating data after the first 64 data points. There *must* be something in the spectrum which *starts and stops* the reconstructed data—something that makes it appear in the first half of the 128 data point array. If the *even* components are identical to the original transform components, it would appear the *odd* components *must* contain the information about *starting and stopping in the front half*; but, do these *odd* frequency components make this spectrum unusable as a spectrum analyzer?]

Run FFT09.02 again and select illustration 3. This time we make the array *four times* the length of the data being analyzed—we are now analyzing the *data* at 1/4 Hz intervals. If we scan the spectrum the amplitude varies even less than before; but, also, the garbage is beginning to look coherent...hey guys, is that a light I see at the end of the tunnel?

Enough of this—let's really cut to the chase. Run illustration 5 and now you will see that the garbage actually has a well-defined structure (Fig. 9.3). These bumps are (to a good approximation) the absolute value

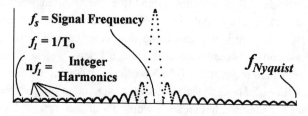

Figure 9.3 - 1/16 Hz Transform

of a *sinc function*. What's really going on here?

We all see that this sinc function appears because we have packed the aft end of the array with zeros, but some suppose it's generated by an

The Spectrum Analyzer

illegitimate(?) configuration. Some, however, may see there's nothing really wrong with this configuration, and suspect the sinc function is just the inherent *bandpass* characteristic of Fourier analysis. Still others may see that, when we view a spectrum via the conventional transform, we're sort of looking through a picket fence, and when we analyze sinusoids that exactly match the orthogonal Fourier components the sinc function just happens to be hidden behind the pickets; however, when harmonics of the input function do not match the Fourier components, the sinc function appears in the spectrum. Consequently, when we analyze the data with 1/2 (or 1/4, etc.) Hz components, the sinc function can no longer be hidden, and as we increase the frequency domain resolution via fractional frequencies, we begin to see the real nature of the transformed function (i.e., we see between the pickets of the fence).

There's an obvious consideration here that's virtually invisible to the uninitiated. The data array that we analyze must always be limited to a finite number of data points—real data must *always* have a beginning and an end. If we simply follow instructions, and do things as we are told, we may indeed miss the significance of this; however, if we do things differently (e.g., we put our sampled data into an oversized array), we begin to see a sinc function spectrum (implying that our data has a beginning and an end). Let's try to make this clearer.

9.3 THE DREADED SINC FUNCTION

Note: You may have recognized that, while we're ostensibly illustrating how the FFT is applied, we spend a lot of time kicking around fundamental phenomena underlying this technology. We introduce the sinc function here, but we will return to this profound little gizmo repeatedly in the following chapters.

Let's cut directly to the chase—what the heck's a sinc function? A sinc function is defined as Sin(x)/x. That is, it's a sine wave divided by its own argument so the amplitude decreases as the argument increases. Interestingly, although the *sine* function starts with a value of zero, the value of Sin(x)/x starts at 1.0 (we explained why this happens back on *p.* 6 at the end of Chapter 1), but there are

Figure 9.4 - The Sinc Function

less obvious attributes we should discuss. Technically, Sin(x)/x is the "Saw" function [i.e., Sa(x)]—the *sinc function* is really Sin(πx)/πx. The distinction arises since the argument of a sine function must be an angle, so the introduction of π changes the nature of x. That is, the introduction of π implies *radians* (x is then a *pure* number)—otherwise x itself must be an angle. So, if we define our function as Sin(πx)/|πx| (where |πx| is the magnitude only) we are very close to what is shown in Fig. 9.4. On the other hand, if we write Sin(πx)/πx we are technically saying that the sine of x is divided by a *rotational entity*—an angle—a complex quantity. Table 9.1 compares several variations on this theme [note that, in complex quantity versions, this function dwells in the third/fourth quadrants].

x	Sin(πx)/x	Sin(πx)/|πx|	Sin(πx)/πx	Sin(x)/x
0.0	+3.141	+1.000	+1.000 -J0.000	+1.000 -J0.000
0.2	+2.939	+0.935	+0.757 -J0.550	+0.974 -J0.197
0.4	+2.378	+0.757	+0.234 -J0.720	+0.897 -J0.379
0.6	+1.585	+0.505	-0.156 -J0.480	+0.777 -J0.531
0.8	+0.735	+0.234	-0.189 -J0.137	+0.625 -J0.643
1.0	-0.000	-0.000	+0.000 -J0.000	+0.455 -J0.708
1.2	-0.490	-0.156	+0.126 -J0.092	+0.281 -J0.724
1.4	-0.679	-0.216	+0.067 -J0.206	+0.120 -J0.694
1.6	-0.594	-0.189	-0.058 -J0.180	-0.018 -J0.624
1.8	-0.327	-0.104	-0.084 -J0.061	-0.123 -J0.527
2.0	+0.000	+0.000	+0.000 -J0.000	-0.189 -J0.413
2.2	+0.267	+0.085	+0.069 -J0.050	-0.216 -J0.297
2.4	+0.396	+0.126	+0.039 -J0.120	-0.208 -J0.190
2.6	+0.366	+0.116	-0.036 -J0.111	-0.170 -J0.102
2.8	+0.210	+0.067	-0.054 -J0.039	-0.113 -J0.040
3.0	+0.000	+0.000	-0.000 -J0.000	-0.047 -J0.007
3.2	-0.184	-0.058	+0.047 -J0.034	+0.018 -J0.001

Table 9.1

So, how do these sinc functions get into the spectrums of fractional frequency transforms? More importantly, is it really contained (but hidden) in the spectrums of data we obtain with conventional transforms? These are important questions for they are, literally, the difference between seeing the world through picket fences and seeing the world as it really is. If we can find out how the sinc function gets into the Fourier transform in the first place, we can probably tell if it's always present (when data starts and stops). Let's start by considering the archetypical *start/stop* function.

The Spectrum Analyzer

9.4 THE UBIQUITOUS SINC FUNCTION

When we introduced the concept that any complex waveshape could be created by summing sinusoids, we used the example of a square wave built from odd harmonics (whose amplitudes were inversely proportional to their harmonic number).

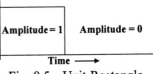

Fig. 9.5 - Unit Rectangle

Let's go back and look at that square wave from a fractional frequency point of view. [Note that, if the *on* time equals the *off* time we allow the wave is *square*—otherwise we refer to it as a *rectangle* wave.] We modify program FFT09.02 as follows:

```
400 REM GENERATE RECTANGLE FUNCTION
.
.
412 S(I) = 1
.
```

This generates a function of constant value 1.0. We will also need to change the scale factor for displaying the data:

```
350 CLS : X0 = 50: Y0 = 300: XSF = 500 / Q2: YSF = 250 * Q / QI
```

If we now run the program (1/16 Hz analysis) we will get the display shown in Fig. 9.6—once again, the absolute value of a sinc function.

Now, it's intriguing that the transform of a rectangle wave is a sinc function—rectangle functions do little more than start and stop (i.e., if we multiply a second function by this rectangle function, the rectangle serves only to start and stop the second function). We can almost see where this argument is going—we can almost see a distant castle beginning to appear through the slowly clearing fog. We can help the fog along by considering another of those fundamental notions that will come up repeatedly throughout this book—one we have already introduced—the phenomenon of modulation.

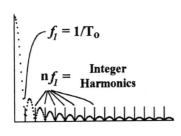

Fig. 9.6 - Rectangle Xform

9.5 MODULATION

As we discussed back in Chapter 3 (Section 3.2, p.21), if we multiply one sinusoids by another we get a unique product called *sidebands* (see fig. right). If we multiply a high frequency sinusoid by a function that's composed of a summation of sinusoids (e.g., a rectangle wave), the modulated sinusoid will be replaced with sideband pairs for every harmonic of the *modulating* wave. That is, if we modulate (i.e., multiply) a sinusoid by the rectangle function described on the previous page, the sinc function spectrum will replace the sinusoid.[2]

If, then, we modulate some complicated waveshape (containing multiple harmonics) with a rectangle function, we must surely find that every harmonic of the modulated waveform is replaced by sideband components equal to the harmonics of the modulating waveform. That is, every harmonic of the *modulated* wave will be started and stopped by the rectangle function, and therefore replaced by the sidebands of the modulating wave. So, it's apparent that, in our fractional frequency transform, this is what we have done. We have multiplied the signal by a rectangle function (in the *time* domain, see Fig. 9.8), and the *sinc function spectrum* must appear about the harmonics of the input function in the frequency domain.

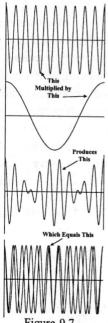

Figure 9.7 - Modulation

Clearly, then, whenever we start and stop data (as in a fractional frequency transform) we are effectively multiplying by a rectangle function, and the sinc function must always appear in the resulting spectrum. But wait—this is true whether we analyze with fractional frequencies or not! When we start and stop data, mathematically we are multiplying by a rectangle function, and the sinc function spectrum will appear about the harmonics of the sampled data function (Fig. 9.3). In a sense, these sinc function sidebands *are* the *frequency domain* start and stop...the beginning and end of the sampled data interval.

[Note: If the function we sample is more complicated than a single

---

[2] We should note that, if the modulating function contains a constant term (i.e., a DC component), the sinusoid being modulated *will not disappear from the product.*

sinusoid (i.e., if it's composed of harmonic components) and every harmonic has its own set of sinc function sidebands, then the resulting spectrum will not be a simple sinc function. In general, spectrums will be quite complicated (since all the sinc function sidebands fall on top of each other), and what we *see* for any harmonic component will be the summation of this convoluted mess.][3]

Okay, the spectrums of *all* data having a beginning and an end will be *convolved* with a sinc function spectrum; however, in the *conventional* FFT, the components of this sinc function are *hidden*. That is, if the harmonic components of the input signal fit exactly into the domain of the data, the sinc function components will always be zero (Figs. 9.3 and 9.6). Indeed, that is precisely

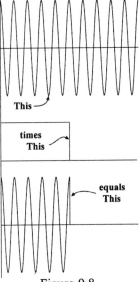

Figure 9.8

why we select these specific components. If, then, we digitize some periodic function (e.g., a sinusoid), and our data interval is not an exact multiple of it's period, we will encounter the spectrum of the sinc function (Fig. 9.9), for it doesn't matter whether the analyzing function or the function analyzed results in non-orthogonality. This, then, is the phenomenon we have discovered in our attempts to design a spectrum analyzer. This isn't just a mathematical gimmick—the sinc function is the frequency domain representation of data that starts and stops!

So, the sinc function must indeed be ubiquitous—it's present in the spectrum of all data that begins and ends—it's present in the spectrum of all real data we measure. Not only is it present—it will be visible if we look for it. It will rear its dreaded head every time we measure data that contains *non-orthogonal* sinusoids on the data interval. That is, with real data, it will be present all the time)!

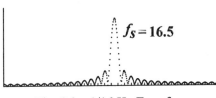

Fig. 9.9 - 1/16 Hz Transform

---

[3] If you trace this out, step-by-step, you will see it's identical, mathematically, to the procedure we called *convolution* in the last chapter of the first book.

9.6 SINC FUNCTION AND RESOLUTION

Until this chapter we have limited our Fourier analysis to *orthogonal* sinusoids. When we interface with the real world, however, we find the sinusoidal components within any signal will seldom be perfectly orthogonal with the time domain interval. We can't see these real world harmonics with fidelity unless we use a fractional frequency transform, and when we employ fractional frequencies the dreaded sinc function appears. Now, using *integer fraction* frequencies doesn't really violate the orthogonality requirement—it's equivalent to looking at the input signal on a larger domain—that's a major point we're trying to make here.

In practical spectrum analysis there are two more or less obvious problems created by the sinc function—*resolution*, and *side-lobe interference*. As shown in Figure 9.3, the sinc function *broadens* the components of the input signal, making it impossible to distinguish two components that are close together. From the *similarity theorem*, of course, we know that if we expand the rectangle (i.e., we sample the data for a longer interval), the sinc function becomes narrower. The way to resolve components that are close together, then, is by sampling the data for longer intervals. We include this provision in version 2 of the spectrum analyzer (FFT9.02).

Fig. 9.10 - Similarity

9.7 INTERFERENCE OF THE SIDE LOBES

Our second problem with the sinc function is this: the sinc function side-bands are of non-negligible amplitudes. These *side lobes* ride on the signal components and wreak havoc on *quantitative* analysis. Now, in a practically useful instrument we expect to resolve signals more than 100 db down. To see the severity of this effect we must display the spectrum in the way it's most often displayed on spectrum analyzers—as the logarithm of amplitude *ratios* (i.e., in terms of deci-bels). We use a *db* scale display in FFT09.02 and, in Fig. 9.11,

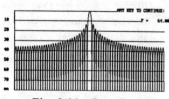

Fig. 9.11 - Log Scale

The Spectrum Analyzer

show the effect of the side lobes.[4] This, obviously, is unacceptable.

Now, when we studied filters in Chapter 3, we noted a similar phenomenon (in the time domain) for a truncated spectrum (i.e., Gibbs' phenomenon). In that particular case, we noted that we could fool the system by using a Gaussian filter. That is, by gently *rolling-off* the harmonics (rather than harshly truncating them), we could make Gibbs go away. So, if we start and stop the *time domain* function gradually, can we make the *sinc function spectrum* go away?

The answer, of course, is yes. By employing *weighting functions*, we can greatly attenuate this effect. The sinusoid we hope to analyze is shown at the top of Fig. 9.12. Since it starts and stops, it *must* possess sinc function sidebands; however, if we multiply this signal by

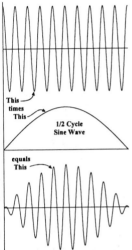

Fig. 9.12 - Weighting

some function that starts and stops gradually, then *its* spectrum will modulate the (harmonics of the) input signal. For example, in Fig. 9.12 we multiply the input signal with a half cycle of sine wave, and indeed, if we run this example in FFT09.02 we find the side lobes *are* attenuated. Several weighting functions are provided in FFT09.02 [i.e., $Sin^2(x)$, $Sin^4(x)$, etc.] with impressive results (see Figs. 9.13 and 9.14) but $Sin^{2N}(x)$ is not the only possibility[5].

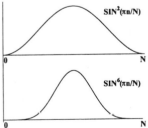

Fig. 9.13 - Weighting Functions

FFT09.02 is designed to allow experimentation with fractional frequencies, data interval, and weighting functions. The two characteristics we are most interested in are *component separation* and *fidelity of quantitative results*. You will find that the variables of this program are interactive, and the use of a *weighting function* doesn't mean fractional frequencies are unnecessary (try holding amplitudes accurate to 0.01 db for example).

---

[4]There is an incredible amount of confusion about the deci-bel. In spite of what you have heard, there's only *one* deci-bel (see Appendix 9.3).

[5]See Ch. 4 of Smith and Smith's *Handbook of Real Time Fast Fourier Transforms*, IEEE Press. They discuss 13 different weighting functions (23 illustrations).

9.8 CONCLUDING OBSERVATIONS

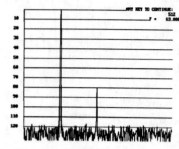

Fig. 9.14 - -80 db Component

It just *might* be possible to produce improvements over the old analog instruments. Don't be misled by the slowness of these programs—BASIC is notoriously slow![6] With memory so cheap, we could digitize data for long intervals, so frequency resolution is *potentially* better than the old instruments. Also, combining fractional frequency analysis with the weighting functions begins to bring quantitative data in line (even with the limited program presented here we can approach 0.01 db). The software can be speeded up tremendously by simply getting out of BASIC, but there are a great many other ways to speed things up. For example, the FFT is well suited for multiprocessor application. The *odd* and *even* data points separate easily and can be processed simultaneously—the *stages of computation* are also well-suited for *pipeline* architectures.

In our studies up until now we have made a great deal over the need for orthogonal sinusoids—integer multiple frequencies with *average values* of zero over the domain of definition—we have flatly *insisted* on orthogonality of the basis function components (see Chapter 2 of *Understanding the FFT* and Sections 3.4 in this book). When we consider that there's no difference in the analysis of a partially filled data array (i.e., one that has been packed with trailing zeros) and simply taking the transform of a full array using *fractional frequencies*, then fractional frequencies don't seem so bad. We recognize, of course, that on a larger domain (i.e., one packed with zeros), these fractional frequencies *are* of zero average value—they are orthogonal. On this larger domain they only detect (reveal) the condition that the function being analyzed has a beginning and an end. So, then, it would appear that we may not only use integer multiple frequencies, but also frequencies that are the reciprocals of the integers (i.e., 1/2, 1/4, 1/8, etc.) in our DFT analysis.

---

[6]"Chip Sets" are available from major manufacturers which run a 1024 point FFT in typically about 1 to 2 ms (i.e., .001 to .002 sec.).

CHAPTER 10

ABOUT SAMPLING THEORY

10.0 INTRODUCTION

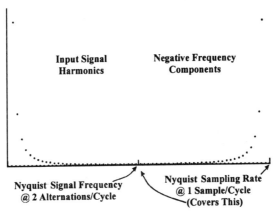

Figure 10.1 - Nyquist Sampling Relationships

Since practical systems have finite bandwidths there will always be some highest frequency of interest. As you may already know, to properly digitize a signal we must sample at *twice* the highest frequency contained in that signal (according to Claude Shannon's sampling theory).[1] This is the Nyquist sampling rate as everyone knows; but, *why* is the ratio of 2:1 so sacred? Why not 1:1...or 4:1... or 10:1? Now, it's reasonably obvious that, if we take a sufficiently large number of samples (say 8 to 16 for each cycle), the original signal will be faithfully captured; but, why is *two* considered necessary and sufficient? *This* is a puzzle worthy of a little effort, and since we sincerely believe this is the sort of thing we're good at, we really can't just turn away....

---

[1] In 1948 Claude E. Shannon published a two-part paper which is now regarded as the foundation of modern information theory. In his 1949 paper *Communication in the Presence of Noise*, he called the critical sampling interval the *Nyquist Interval* and the name stuck. Why *Nyquist*? Harry Nyquist, working for AT&T in 1924, published his paper, *Certain Factors Affecting Telegraph Speed*, in which he defined line speed as one half the number of signal elements which can be sent per second. E.T. Whittaker, however, may claim priority, publishing his work in 1915.

Okay, our reputation is at stake, so we will claw and scratch for *any* hand-hold to get started. Now, Fig. 10.2 shows two *alternations* to each *cycle* of a sinusoid, and it's apparent that, if we hope to capture every little bump in the signal, we must have one sample for each alternation (at least). That, of course, would be two samples per cycle of signal.

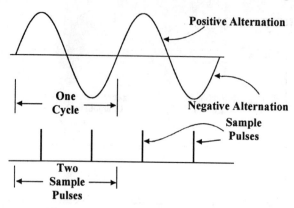

Figure 10.2 - Nyquist Samples per Cycle

Let's clear a hole in the fog here: in discussing this phenomenon we will sometimes refer to the highest frequency component in the signal as the Nyquist frequency *while the sampling rate will be referred to as the* Nyquist sampling rate. *This confusing terminology comes about simply because of the way we specify signal frequency (as opposed to sampling rate). The "Hertz" implies a number of full sinusoids (two alternations in a cycle) per unit time while* Samples per Second *implies a number of single events (samples) per unit time. If we were to define the frequency of sinusoids by the* number of alternations *per unit time, then the 2:1 ratio would become 1:1. The point is this: in a sense the sampling rate is actually the same thing as the Nyquist signal frequency—it's just that we only count one sample per "cycle" of sampling rate, but we count two alternations for a* cycle of signal *(compare Figure 10.1 to Figure 10.2).*

So, it appears there's nothing mysterious about the Nyquist sampling criterion of 2:1...but there's a bit more to it than just this two alternations vs. one sample per cycle. Surely 2:1 is imperative if we hope to capture the highest frequencies, but is 2:1 *enough*? Can we really (in the practical world) capture and *reproduce* a high fidelity replica of any signal at 2:1? Let's investigate some of the sampling rate implications that may not be immediately obvious.

Sampling

10.1 NEAR-NYQUIST SAMPLING

Sampling at the Nyquist rate provides a sort of *bare bones* minimum of information about the signal, and if things work out as shown

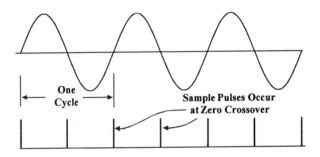

Figure 10.3 - Out-of-Phase Sampling

in Fig. 10.2, we will capture every fluctuation in the signal; however, what if the signal and sampling pulses are shifted in phase? Suppose things are as shown in Fig. 10.3? Here the digitized data will indicate no signal when, in fact, a perfectly good sinusoid is present at the input.

Actually, the situation shown in Fig. 10.3 is of little practical consequence. In a practical system we will never have a sinusoid of significant amplitude exactly at the Nyquist frequency. Our discussion of filters in Chapter 6 makes it obvious that, if there are to be no components above the Nyquist, then (due to the finite *roll-off rate* in physically realizable filters) the harmonics must already be of negligible amplitude at the Nyquist frequency. What we must worry about are the components just below the Nyquist—here we *will* have harmonics of significant amplitude. So, what happens to sampled data in this part of the spectrum?

Once again, let's use a short computer program:

```
10 REM ******************************************
11 REM *              NYQUIST TEST 1            *
12 REM ******************************************
20 PI = 3.141592653589793#:F1 = 0.9 ' F1 = % NYQUIST
30 FOR N = 0 TO 20
32 X = N*F1*PI ' SOLVE FOR ARGUMENT
40 Y = COS(X) ' SOLVE FOR COSINE
48 PRINT N,:PRINT USING "##.#####";Y
50 NEXT ' DO THIS 20 TIMES
60 END
```

Note: The basic scheme here is to sample a sinusoid at rates near the Nyquist, but since we are only concerned with the output of the A/D converter, we need only generate numerical values of the sine function at the sampling points.

If F1 had been set to 1.0 in line 20 we would simply obtain the alternating values of +1 and -1 [i.e. $Cos(N\pi)$, where N = integer values from 0 to 20]. This would be the Nyquist condition—two samples per cycle. Since F1 = 0.9 however, the values digitized slip slowly out of phase...and then back in again. Figure 10.4 illustrates how this happens graphically, and if we take a pencil and connect every other value, we trace out the envelopes of two interleaved sinusoids.

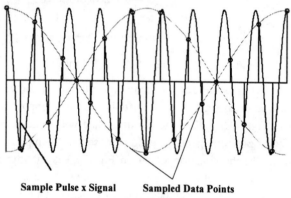

Sample Pulse x Signal   Sampled Data Points
Figure 10.4 - Data Sample Points

This, then, is our *near-Nyquist* digitized signal—and our first observation is that **the digitized pattern does *not* look like the analog signal we had to digitize**—there was no modulation on the sinusoid we digitized! One would think the digitized data should look like the input; nonetheless, Claude and Harry say this is valid digital data. (?)

It's reasonably obvious why sampled sinusoids entrain sinusoidal envelopes, but there's more to this than the obvious. You may recall seeing similar figures (*pp*. 19-20 and *p*. 100). The similarity between Fig. 10.4 and the result of multiplying two sinusoids is striking, but we're not multiplying sinusoids here—we're sampling data. We may visualize *sampling* as the mathematical *product* of the input signal and a series of impulses (each of unit amplitude), it being clear the sampling pulse train, having a value of zero between impulses, yields only values equal to the amplitude of the signal at the time of each impulse.

Sampling

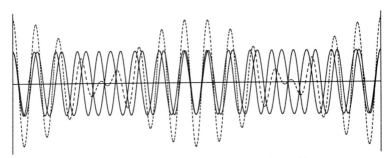

Figure 10.5 - Summation of Two Sinusoids

Sinusoids are pretty simple things when you get to know them (never met a sinusoid I didn't like); so, when you see a distinctive pattern (e.g., Figs. 10.4 and 10.5) you can't help but suspect that, under that flimsy disguise, it's the same guy (or at least a close relative).[2] Now, if we had digitized the sum of the two sinusoids shown in Fig. 10.5 we would *expect* the result to look like Fig. 10.4; but, we only digitized a single sinusoid! Let's run another experiment.

Instead of setting F1 to 0.9, let's make it 1.1. This is an illegal condition according to Claude and Harry—it's 10% *above* the Nyquist limit—but let's just see what happens when we break this rule. (*We can make life easier here with a little modification to our computer program:*)

```
20 PI = 3.14159265358979#: F1 = .9: F2 = 1.1' F1 & F2 = % NYQUIST
30 FOR N = 0 TO 20
32 X = N * PI: X1 = X * F1: X2 = X * F2' SOLVE FOR ARGUMENTS
40 Y1 = COS(X1): Y2 = COS(X2)' GET COSINES
48 PRINT N, : PRINT USING "##.#####"; Y1,Y2
50 NEXT N
```

Table 10.1 (next page) shows the output of this routine, comparing the two digitized sinusoids (one 10% below Nyquist and the other 10% above). The two printouts are virtually identical—there appears to be no difference between the digitized outputs of a sinusoid 10% above Nyquist and one 10% below. Furthermore, if you play with this routine just a little you will find that *any* two pairs of signals spaced equally above and below the Nyquist give identical digitized data.

---

[2]Fig. 10.5 is a great picture—it's worth at least five or six hundred words (on a scale of zero to a thousand words/picture), showing explicitly why the sum of two sinusoids generate the same sinusoidal envelope as two multiplied sinusoids.

N	F1 = 0.9	F1 = 1.1
0	1.00000	1.00000
1	-0.95106	-0.95106
2	0.80902	0.80902
3	-0.58778	-0.58778
4	0.30902	0.30902
5	-0.00000	0.00000
6	-0.30902	-0.30902
7	0.58778	0.58779
8	-0.80902	-0.80902
9	0.95106	0.95106
10	-1.00000	-1.00000
11	0.95106	0.95106
12	-0.80902	-0.80902
13	0.58779	0.58778
14	-0.30902	-0.30902
15	0.00000	0.00000
16	0.30902	0.30902
17	-0.58778	-0.58779
18	0.80902	0.80902
19	-0.95106	-0.95106
20	1.00000	1.00000

Table 10.1 - Comparison of 10% Above Nyquist with 10% Below

We see a little more of Claude's problem here. If sinusoids spaced equally above and below the Nyquist give identical digitized results, there's no way to distinguish between them later. If we violate the Nyquist criterion then, and allow frequencies both above and below the Nyquist, the digitized resultant will be a hopelessly confused summation of the two.[3] We can avoid this morass by band-limiting the signal (the one to be digitized) to frequencies below the Nyquist (i.e., below 1/2 the sampling rate), but we will still have the beat frequency modulation phenomenon for frequencies near the Nyquist.

10.2 NECESSITY AND SUFFICIENCY

Our question is why conventional theory says a sampling rate of twice the highest signal frequency is necessary *and* sufficient. We have just seen that if the sampling rate is *not* twice the highest frequency *aliasing* will occur. This makes the Nyquist criterion a little more significant than our two alternations per cycle observation—but the second part of the question concerns whether a 2:1 ratio is *sufficient* to provide distortion free digitization (and reproduction) of the data. If you play with

---

[3] Summing sinusoids from *above* the Nyquist into the digitized data is termed *aliasing*, and obviously compromises the sampled data.

Sampling

the digitization routine given above (i.e., Nyquist Test 1), you will find that sampling ratios of less than about 8:1 *frequently* give poor replicas of the input sinusoids, and we must wonder if this sort of thing is acceptable. Worse yet, the presence of a *beat frequency modulation* envelope hints there may be sinusoids in the digitized data that were not in the original signal—that this *modulation* is really *intermodulation distortion*.... So, is Shannon's 2:1 ratio really sufficient?

10.3 SAMPLING PULSE TRAIN MODULATION

Again we return to the topic of modulation. As we well know, when we multiply sinusoids, the original sinusoids disappear and two new side-band sinusoids are produced (see Ch. 3). Furthermore, the frequencies of the sidebands are equal to the sum and difference of the frequencies that were multiplied together. *[We're beginning to see that modulation (frequency domain convolution) is a profound phenomenon—one that's critical to understanding the world around us.]*

Now, we're not modulating sinusoids when we digitize a signal—we're modulating *sample pulses* (see Figure 10.2 and 10.4), and we need to look at these sampling pulses in terms of Fourier analysis....

10.4.1 SAMPLING PULSES

First of all, we have sort of *created* this sampling pulse! In fact, *this* sample pulse has no physical existence. Its conceptual existence comes about when we *back-through* the A/D conversion process, for in the process of digitizing a signal we create *perfect* numbers which have no extension in time or space. The digitized data *points* are mathematical points that would be created *if* we multiplied the analog signal by a train of impulses (of unit amplitude and *infinitesimal* width). *[The mechanics of A/D conversion are just technicalities as far as this line of reasoning is concerned.]*[4] So, we have deliberately created a fictitious *sampling pulse*, which is the mathematical equivalent of what *would have* created our sampled data, and we are now going to investigate its frequency spectrum.

---

[4]The *data* extends in time and/or space, and the conversion process must include this extension to some extent, but the numbers know nothing of this. We may use them in mathematical developments that assume they are extended in time, but the numbers know of that either!

10.4.2 IMPULSE SPECTRUM

As you recall, when we illustrate the fundamental scheme of the Fourier series (by summing odd numbered harmonics with amplitudes inversely proportional to their harmonic numbers), our efforts are rewarded by the gradual, almost magical appearance of a beautifully simple function—a square wave. Now, if we sum the simplest series of sinusoids possible—a series of cosine waves with no variation in amplitude—no leaving out odd or even harmonics—just a simple summation of cosine waves—we get a single data point (i.e., an impulse) *followed by a flat line of perfectly zero amplitude.*

So, the question that comes to mind is, how does the summation of our simple, smooth, continuous, sinusoids create the most discontinuous wave shape imaginable? This is another of those fundamental phenomena that will turn out to be more important than you might think; so, let's use another computer routine to investigate what happens here:

```
'    ****************************************************
'    *     APPLICATIONS 10.2 - IMPULSE FUNCTION TEST     *
'    ****************************************************
12 INPUT "NUMBER OF TERMS AS 2^N"; N 'NUMBER OF SINUSOIDS TO SUM
20 Q = 2^(N+4)' MULTIPLES OF 16 FREQUENCY COMPONENTS
22 PI = 3.14159265358# 'NO IDEA BUT CAN'T STOP NOW...
30 FOR I = 0 TO 2 * PI STEP PI / 8 'COMPUTE EACH DATA POINT
32 Y = 0 'INITIALIZE ACCUMULATOR
40 FOR J = 0 TO Q-1: Y = Y + COS(I * J): NEXT 'SUM ALL HARMONICS
50 PRINT USING "####.#####"; Y 'PRINT DATA POINT
60 NEXT I 'DO NEXT DATA POINT
62 INPUT A$' WAIT HERE, TRUSTY 286...
```

When you run this routine you will notice that the single point at the origin has a value equal to the number of sinusoids summed (Table 10.2). The reason is fairly obvious—the value of a cosine is 1.0 when the argument is zero. If we sum Q cosine waves, which all start with an argument of zero (regardless of their frequency), they will all have an initial value of 1.0; therefore, the initial data point *must* have a value of Q. Table 10.2 shows the value of each of the harmonic components (i.e., F00, F01, etc.) *at the data sample times* (T00, T01, etc.). [It took a lot of time to create this picture so I would appreciate it if you perused it carefully.]

So, the first data point is 16.000, but how did all the other data points get to be *exactly* zero? Let's step through the operation of this routine and see what really happens. In line 40 we accumulate Cos(I*J) for

Sampling

	T00	T01	T02	T03	T04	T05	T06	T07	T08
F00	+1.000	+1.000	+1.000	+1.000	+1.000	+1.000	+1.000	+1.000	+1.00
F01	+1.000	+0.924	+0.707	+0.383	-0.000	-0.383	-0.707	-0.924	-1.00
F02	+1.000	+0.707	-0.000	-0.707	-1.000	-0.707	+0.000	+0.707	+1.00
F03	+1.000	+0.383	-0.707	-0.924	+0.000	+0.924	+0.707	-0.383	-1.00
F04	+1.000	-0.000	-1.000	+0.000	+1.000	-0.000	-1.000	+0.000	+1.00
F05	+1.000	-0.383	-0.707	+0.924	-0.000	-0.924	+0.707	+0.383	-1.00
F06	+1.000	-0.707	+0.000	+0.707	-1.000	+0.707	-0.000	-0.707	+1.00
F07	+1.000	-0.924	+0.707	-0.383	+0.000	+0.383	-0.707	+0.924	-1.00
F08	+1.000	-1.000	+1.000	-1.000	+1.000	-1.000	+1.000	-1.000	+1.00
F09	+1.000	-0.924	+0.707	-0.383	-0.000	+0.383	-0.707	+0.924	-1.00
F10	+1.000	-0.707	-0.000	+0.707	-1.000	+0.707	+0.000	-0.707	+1.00
F11	+1.000	-0.383	-0.707	+0.924	-0.000	-0.924	+0.707	+0.383	-1.00
F12	+1.000	+0.000	-1.000	-0.000	+1.000	+0.000	-1.000	-0.000	+1.00
F13	+1.000	+0.383	-0.707	-0.924	-0.000	+0.924	+0.707	-0.383	-1.00
F14	+1.000	+0.707	+0.000	-0.707	-1.000	-0.707	-0.000	+0.707	+1.00
F15	+1.000	+0.924	+0.707	+0.383	+0.000	-0.383	-0.707	-0.924	-1.00
	16.000	0.000	0.000	0.000	0.000	0.000	0.000	0.000	0.00

TABLE 10.2 - Summation of Equal Amplitude Sinusoids

all Q harmonics. When I = 0 (i.e., the initial data point) Cos(I*J) will be 1.0 regardless of the value of J (J = harmonic number). The next data point, however, gives I a value of $\pi/8$. Now, as J steps through each of the 16 harmonics, I*J steps through 2π radians, and Cos(I*J) will give a sinusoidal pattern of values—the summation of which is zero (see column T01). At the next data point the value of I will be $2\pi/8$ and we will only have to step through 8 harmonic numbers to get a null summation (which we will do twice). Table 10.2 shows the summation of 16 harmonics (F00 through F15) for the first 8 data points (T00 through T08). This is simple but important—these summations simply *must* come out to zero. [You recognize, of course, that we're playing with loaded dice again; *but, this is* the very configuration we use in the DFT and FFT. Now, while we are playing with loaded dice in the above illustration, it's important to understand that this phenomenon (i.e., a large number of sinusoids summing to zero under conditions of symmetry) occurs *naturally* (i.e., even when the dice are *not* loaded).]

For now, we should note that these sinusoids will *fail to yield a zero summation* whenever I = $16\pi N_I/8$ [where N_I is any integer (including zero obviously)]. At these points Cos(I*J) = 1.0 regardless of the integer value of J (as we said J represents harmonic numbers, or frequency); however, the domain of our consideration extends only from I = 0 to one data point less than I = 2*PI. *Note: The cosine waves we have been* discussing here, all of equal amplitude and constant phase, represent the transform *of the impulse function, of course. If we* take *the transform of an impulse we will obtain a perfectly flat spectrum of cosine waves.*

So, the impulse function has an amplitude equal to Q (where Q is the number of harmonics in the frequency domain function), but if we divide each harmonic by Q, they will all sum up to unity, and this function will fulfill our requirements for a single *sample pulse*.

10.4.3 FREQUENCY DOMAIN STRETCHING

But this is a single impulse—to model the sampling process we need a "train" of sampling pulses. We can get to the sampling pulse train in various ways, but let's use the *frequency domain stretching* theorem here.[5] So, if we take an *impulse function spectrum* (of equal amplitude harmonics—Table 10.2) and *stretch it out* by placing zeros between each of the harmonic components, we get an interesting result on reconstruction of the time based waveform. If we place zeros between all the harmonics (equivalent to multiplying all the frequencies by two) we will have only *even* numbered harmonics (with all the *odd* harmonics equal to zero).[6] You

	T00	T01	T02	T03	T04	T05	T06	T07	T08
F00	+1.000	+1.000	+1.000	+1.000	+1.000	+1.000	+1.000	+1.000	+1.00
F01	+0.000	+0.000	+0.000	+0.000	-0.000	-0.000	-0.000	-0.000	-0.00
F02	+1.000	+0.707	-0.000	-0.707	-1.000	-0.707	+0.000	+0.707	+1.00
F03	+0.000	+0.000	-0.000	-0.000	+0.000	+0.000	+0.000	-0.000	-0.00
F04	+1.000	-0.000	-1.000	+0.000	+1.000	-0.000	-1.000	+0.000	+1.00
F05	+0.000	-0.000	-0.000	+0.000	-0.000	-0.000	+0.000	+0.000	-0.00
F06	+1.000	-0.707	+0.000	+0.707	-1.000	+0.707	-0.000	-0.707	+1.00
F07	+0.000	-0.000	+0.000	-0.000	+0.000	+0.000	-0.000	+0.000	-0.00
F08	+1.000	-1.000	+1.000	-1.000	+1.000	-1.000	+1.000	-1.000	+1.00
F09	+0.000	-0.000	+0.000	-0.000	-0.000	+0.000	-0.000	+0.000	-0.00
F10	+1.000	-0.707	-0.000	+0.707	-1.000	+0.707	+0.000	-0.707	+1.00
F11	+0.000	-0.000	-0.000	+0.000	-0.000	-0.000	+0.000	+0.000	-0.00
F12	+1.000	+0.000	-1.000	-0.000	+1.000	+0.000	-1.000	-0.000	+1.00
F13	+0.000	-0.000	-0.000	-0.000	-0.000	+0.000	+0.000	-0.000	-0.00
F14	+1.000	+0.707	+0.000	-0.707	-1.000	-0.707	-0.000	+0.707	+1.00
F15	+0.000	+0.000	+0.000	+0.000	+0.000	-0.000	-0.000	-0.000	-0.00
	8.000	0.000	0.000	0.000	0.000	0.000	0.000	0.000	8.00

TABLE 10.3 - Summation of Equal Amplitude Sinusoids

---

[5]The frequency domain stretching theorem will play an important role in Chapter 12 when we change the playback rate of recorded audio.

[6]Note that F01 in the F02 position doesn't generate the same values it produced in the F01 position.

Sampling

recognize this is the *only* way we can stretch the spectrum with our *DFT loaded dice* scheme. This result is illustrated in Table 10.3, where you will note that a minor miracle takes place—we get a second pulse in the T08 column (which, we recognize, would be halfway through the time domain data array). This is the frequency domain equivalent of the *time-based stretching* we used when we developed the FFT algorithm in *Understanding the FFT*). [Note: A superior intellect would see this phenomenon must happen for *any* function if we stretch its spectrum, but for you guys we will extend this exposition in the next chapter, making it obvious why the time domain function repeats itself when the spectrum is *stretehed*.]

You may have noticed that Table 10.3 and Table 10.2 are actually pictures. They're not the beautiful panoramas of an Ansel Adams, of course...the beauty of these pictures lies in the details...and meaning. They're pictures of sinusoids (of different frequency going down the page) taken at succeeding points in time (going across the page)—they are pictures of the Fourier transform! Table 10.3 is, in fact, just an expanded frequency representation of 10.2. Our sampling times are unchanged between these two pictures, so, in 10.3 we get only half as many sample points in a *cycle* at each frequency. If we generate additional data points between the ones shown in 10.3, however, it becomes identical to 10.2 (ignoring scaling). [Note: It's worth repeating that *all* of the harmonics of the waveform have *doubled* in frequency. If we turn a basso profundo into a chipmunk by playing the record faster, all *the harmonics of that great voice must be scaled proportionally. If we play the record faster, the time domain is compressed and the frequency domain is expanded!*]

Okay, take a sip of coffee here because we're about to reveal a profound truth! In Table 10.3 we have stretched the frequency domain function and so the time domain is compressed—the impulses are closer together. But pulses spaced on shorter time intervals **are** *higher frequency. Compression in time* **is** *expansion in frequency. This most profound of theorems is only profound if we don't think about what we're doing.*

So, stretching the spectrum (i.e., only summing *even* frequency cosines) causes the time domain function to repeat itself (within the same time interval). If we stretch this spectrum again (i.e., we only sum in every fourth harmonic), we will get *four* impulse points. Stretching the spectrum to every eighth harmonic yields eight impulses, and this begins to look very much like a *train* of sampling pulses. In every case, however, the *relative* relationship between the harmonics and impulse rate remains the same! As more pulses occur in a given time, the harmonic *frequencies* increase—the spectrum of this pulse train (made up of identical cosine waves whose frequencies are integer multiples of the fundamental sampling rate) must obviously "stretch."

10.5 SAMPLING PULSE MODULATION (RETURN JUMP)

Okay, sampling is equivalent to multiplying the signal by a string of unit amplitude sample pulses. Now, as we discussed in the previous chapter, if we multiply a signal by a *constant* (for example), we multiply all of the harmonic components by that constant:

$$K_1 f(t) = K_1[A_0 + A_1 Cos(t) + B_1 Sin(t) + A_2 Cos(2t) + B_2 Sin(2t) + ...] \qquad (10.4)$$
$$= K_1 A_0 + K_1 A_1 Cos(t) + K_1 B_1 Sin(t) + K_1 A_2 Cos(2t) + K_1 B_2 Sin(2t) + ... \qquad (10.4A)$$

where: K_1 = constant
$f(t)$ = time domain function

This is fairly obvious—the *relative* amplitudes of the components remain unchanged—they have *all* been scaled up. Suppose now that instead of multiplying our function $f(t)$ by a *constant*, we multiply it by another *function g(t)*. Obviously we now multiply every harmonic of $f(t)$ by $g(t)$:

$$g(t)f(t) = g(t)[A_0 + A_1 Cos(t) + B_1 Sin(t) + A_2 Cos(2t) + B_2 Sin(2t) + ...] \qquad (10.5)$$
$$= g(t)A_0 + \mathbf{g(t)[A_1 Cos(t) + B_1 Sin(t)]} + g(t)[A_2 Cos(2t) + B_2 Sin(2t)] + ... \qquad (10.5A)$$

Take one of the harmonic components from $f(t)$—the $A_1 Cos(t)$ term from Eqn. (10.5A) for example. We know, of course, that $g(t)$ is a function in its own right, and is composed of harmonics:

$$g(t) = U_0 + U_1 Cos(t) + V_1 Sin(t) + U_2 Cos(2t) + V_2 Sin(2t) + ... \qquad (10.6)$$

So, when we multiply $A_1 Cos(t)$ by $g(t)$—$g(t)A_1 Cos(t)$—we multiply it by every harmonic component within $g(t)$:

$$g(t)A_1 Cos(t) = U_0 A_1 Cos(t) + U_1 Cos(t)A_1 Cos(t) + V_1 Sin(t)A_1 Cos(t) +$$
$$U_2 Cos(2t)A_1 Cos(t) + V_2 Sin(2t)A_1 Cos(t) + ... \qquad (10.6A)$$

And, as we know, this replaces the harmonic component of f(t) with *side band* components [i.e., a replica of the spectrum of *g(t)* appears about each and every component of $f(t)$]. This is all pretty obvious, but it's important; so, let's return to our FFT program to illustrate our point.

Sampling

10.6 DATA SAMPLING (RETURN JUMP)

We're going to do something unusual here—we're going to look at something that's normally invisible to *users* of the DFT and FFT—we're going to look at what happens *beyond* the spectrum that's usually displayed. In Section 10.4.3 we saw how a train of impulses (a sampling pulse train) was composed of a spectrum of equal amplitude cosine waves at all integer multiple frequencies of the sampling rate. This spectrum (of the sampling pulses) is normally invisible when we work with sampled data, but in Fig. 10.10, we show the spectrum of a string of impulses. We see that the spectrum is a series of equally-spaced sinusoids (the spacing *between* sinusoids being equal to the frequency of the *sampling rate,* as we just explained for frequency domain stretching).

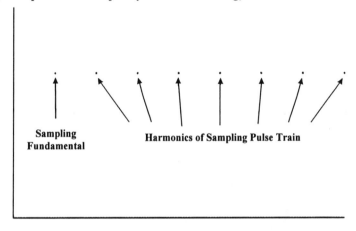

Figure 10.10 - Sampling Pulse Train Spectrum

Okay, if we now use our train of impulses to sample a band-limited triangle wave—the same function we've used extensively before—a triangle wave with the number of harmonics determined by the Nyquist sampling rate (i.e., highest harmonic 1/2 the sampling rate)—and then transform this modulated pulse train...we obtain Fig. 10.11 (next page).

Note that, in the frequency domain displayed, sidebands appear on each and every one of the sampling pulse train's harmonics. This, of course, is the result we described above [Eqns. (10.5A) and (10.6A)]. In Eqn. (10.5A) *f(t)* would represent the sampling pulse train and *g(t)* the modulating wave (a triangle wave in our example). We see exactly what has happened here: each of the harmonics of g(t) must multiply each and every harmonic of f(t). We produce a set of sidebands about each of the sampling pulse train harmonics. We see clearly how sampled data (i.e., modulated sample pulses) relates to modulated sinusoids and, more importantly, how this relates to the sinusoids spaced equally above and below the Nyquist frequency.

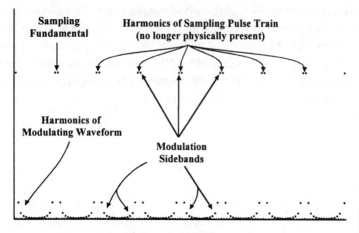

Figure 10.11 - Modulated Pulse Train Spectrum

Again, when we perform a normal DFT/FFT we only consider the portion of Fig. 10.11 which extends from zero up to the fundamental of the sampling rate—as Fig. 10.1 displays. It's now apparent why the negative frequencies are *above* the positive frequencies in a normal DFT—they are the negative frequency sidebands extending downward from the sampling rate fundamental. The positive frequencies are the result of the zero frequency component of the sampling pulse spectrum. We might easily imagine there is another set of negative frequencies *below* the zero frequency shown in Fig. 10.11.

10.7 SUFFICIENCY OF NYQUIST'S CRITERION

Finally, we come back to our original question. Let's suppose we have some utility audio application with a bandwidth of, say, 3200 Hertz. Claude and Harry say a sampling rate of 6400 samples/sec. will get the job done. Suppose then, at some point in time, we have an input signal which

Sampling

is a single sinusoid at 3199 Hz. From what we have just seen, the digitized signal will look something like Fig. 10.4. That is, the negative frequency component, extending downward from the fundamental of the sampling impulse function $f(t)$, will only be 2 Hz away from the single sinusoid of our 3199 Hz signal. Our digitized signal will be (for all practical purposes) a summation of two sinusoids. One at a frequency of 3199 Hz (i.e., the original input signal) and another at 3201 Hz. We can hear both of these, of course, and it will sound awful (like hitting c and c# on a piano—only worse)! If we can filter off the 3201 Hz component, however, everything will then be as it should—the *beat frequency modulation* will disappear and we will only hear the single high-pitched tone—but that's not very realistic. Go back to the filter curves of *p.* 60 and estimate the amount of attenuation provided by the best of those curves when cutoff is exactly at the Nyquist frequency.

The point is this: Shannon says the lower limit for the sampling rate is twice the highest signal frequency. From what we have seen, *if* we can filter off the *negative frequencies* above the Nyquist when we reconstruct the signal, then Shannon's criterion is *sufficient*. Speaking realistically, we must always leave a little room for *overhead* in our system design. Only *if* we can filter off the sidebands extending downward from the first harmonic of the sampling pulse train can we faithfully reproduce the original digitized signal.

Clearly, Shannon's criterion is a theoretical *limit*, and we will seldom obtain that limit. We will look at the practical application of this in detail in the next chapter.

10.8 NYQUIST 2:1 vs. 1:1 SAMPLING

Suppose, instead of sampling the signal with a continuous stream of *positive* impulses, we sample with a string of alternately positive and negative sampling pulses (see Fig. 10.12). If we do this the sample pulse train will have two alternations per cycle, and the frequency of the sample rate will equal the highest frequency in the signal being digitized—there will be

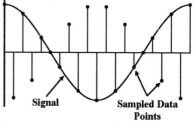

Fig. 10.12 - 1:1 Sampling Rate

only one Nyquist frequency. If every alternation of the pulse train is inverted, a cosine wave at the Nyquist will yield a constant amplitude of sampled data points; however, DC *signal components* (i.e., a constant signal) will yield alternately positive and negative data points (at the Nyquist rate). This is the opposite of what we normally expect from the digitized data; however, it shows clearly what really happens when we digitize a signal. Compare Fig. 10.13 to Fig. 10.1—the negative frequencies now extend downward from the one-and-only Nyquist frequency, and the positive frequencies extend upward.

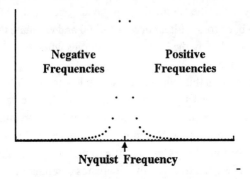

Fig. 10.13 - 1:1 Sampling Spectrum

We note the low frequencies of the sampled signal now appear near the Nyquist frequency while the high frequencies of the sampled signal appear near zero and $2f_N$.

So, in this configuration, the two Nyquist frequencies really *are* the same thing. There's no magic here. There's no loss of information, but neither is there a gain. Still, as with most things we digress upon, we may find a use for this "acorn" later.

CHAPTER 11

DIGITAL HIGH FIDELITY AUDIO

PART I - FREQUENCY RELATED PROBLEMS

In 1877 (only 17 years before the birth of my father) Thomas Edison announced his phonograph. In 1883 Edison noticed that a current would flow (between an incandescent filament and an unheated electrode) through a vacuum. Sir John Fleming used this same effect to build an electronic *valve* (i.e., a vacuum tube *diode* to restrict current flow to one direction only). In 1906 Lee De Forest placed a *grid* between the heated cathode and the anode, thereby allowing a "weak" voltage to control a "strong" current (i.e., to *amplify* a signal). Early amplifiers were not very good, prompting research into lower *distortion* and *higher fidelity*.

The idea of "Hi-Fi" music became popular in the 1940s and '50s (as I recall), but concern for fidelity was already prominent in the '30s. One pictures a progression of technological improvements yielding ever better sound...but this may not be accurate. I recall, in the late '50s, listening to a particular female vocalist (33⅓ rpm stereo vinyl) when I suddenly realized how *real* the sound was. Now, you could always hear the "grooves" of a vinyl recording (in the silence), as well as occasional little "tics," but with a good recording, and a good system, these were *separate* from the music (like someone next to you turning pages at a live performance). If, however, the music *itself* is distorted, we immediately perceive it's not the real thing...and there are those who claim the music recorded on CDs falls into this category. Indeed, there are groups who insist the old vacuum tube amplifiers are better...and refuse to buy CDs...and go to extremes to protect their old vinyl collections.

So, *are* CDs better?

11.1 FREQUENCY RESPONSE

What frequency response do we need for true *high* fidelity? We are told the human ear can hear from *about* 20 Hz to *about* 20,000 Hz; however, when a train rumbles by just outside your motel window, much of what you experience is

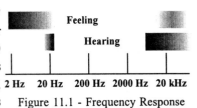

Figure 11.1 - Frequency Response

feeling. This effect is present when a symphony orchestra plays the 1812 Overture (especially when they use real cannons). At the other end of the spectrum we *can* hear frequencies above 20,000 Hz; but these sensations too are more like feeling than hearing. So, how important are these extreme frequencies? How "Hi" do you want your *Hi-Fi*?[1]

When a musical instrument plays a note, it creates a harmonically rich tone. As the instrument plays up and down the scale the relationships of the harmonics change—a trumpet in the high register sounds different from a trumpet in the middle or low register—trumpet players create distinctive tones that are as recognizable as individual voices—even different pianos have different "voices." If we want our sound system to faithfully reproduce the real instruments, we must surely reproduce the major portion of the audible spectrum...but it's virtually impossible to find a compromise specification that will please everyone. If, however, we *faithfully* reproduce the *complete* spectrum, no one can complain...and we can't get it wrong. Regardless of all the opinions about *acceptable compromise*, this must be the ultimate aim of *Hi-Fidelity* sound.

Still, *playing time* (for example) is inversely proportional to bandwidth, and many systems with a frequency response of only 100 to 12,000 Hz sound pretty good; so, what do we really need? Well, the middle A on a piano keyboard is tuned to 440 Hz (this is *concert A*). An *octave* is a 2:1 frequency ratio so the next A going *up* will be 880 Hz, and the next 1760 Hz, etc. Now, if each A is 2:1 in frequency, then each B must also be 2:1, etc. There are 12 tones between octaves (i.e., including *sharps* and *flats* there are 12 keys from A to A), so a little thought reveals that, for an *equal-tempered* scale, each tone must be 1.059463... :1 (i.e., $\sqrt[12]{2}$:1) higher than the previous note. The highest note on a standard keyboard is C at 4186.01 Hz {four octaves up from *middle C* (i.e., the C below *concert A*) which must be tuned to 261.626 Hz (i.e., 3 half-tones up from $A = 220$—$A^\#, B, C)$ times 2^4}. The *thinness* of this highest C tells us there are few harmonics in this tone—its 5th harmonic is 20,930 Hz.[2] Now, 20,930 Hz *is* audible so, if we're trying to *guarantee* realistic sound, shouldn't our *ultimate* system include maybe 20-30% *more* than 20 kHz?

---

[1] Modern CD players spec. the low end at 5 Hz and the high end at 20 kHz.

[2] This is very simplified. Actually, the harmonics of a piano wire are *not* exact integer multiples of the fundamental, causing problems for piano tuners, but causing much greater problems for engineers building *Piano Tuners*. Each note must be tuned to a sort of *weighted average* of the harmonics in each tone.

Digital Audio

On the other end of the keyboard, the lowest note is *A* (4 octaves below *concert A*), at 27.5 Hz. The *over-richness* of harmonics in this last half-octave make these keys infrequently used; nonetheless, we surely hear these frequencies. Again, we *might* limit the low end to 30 Hz...or 25 Hz...but our *ultimate* system will include frequencies below 20 Hz.

11.2 THE PLAYBACK RATE (OVERSAMPLING)

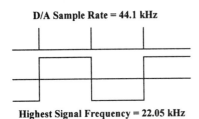

Figure 11.4 - Nyquist Frequency

Okay, we come (finally) to the meat of this chapter, which is to explore FFT applications. We set the frequency response of our hypothetical sound system at 5 to 20,000 Hz. Now, we will need a little *extra* bandwidth (for filter rolloff, etc.); so, for *storing* the data on the CD, we will select a data rate of, say, 44.1 kHz *!-) Now, from Fig. 11.4, this is definitely equal to a Nyquist signal frequency of 22.05 kHz.* We know (Chapter 10) there will be beat-frequency modulation near the Nyquist, implying unwanted sinusoids *above* the Nyquist. Furthermore, from what we know about filters, rolling off the passband between 20 kHz and 22.05 kHz will be difficult.[3]

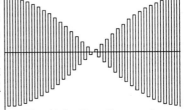

Fig. 11.5 - Beat-Frequencies

The way we handle this is by *oversampling*. Oversampling refers to the technique of introducing *extra data words* into the data stream. For example, we might generate 2:1 oversampling by summing each word with its successor, dividing by two, and inserting the resultant between the digitized data points (dotted lines Fig. 11.6). By inserting *average values* between data samples, we generate twice as many steps per unit time (i.e., the *data rate*, or *sample rate* is doubled to 88.2 kHz).

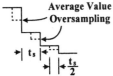

Fig. 11.6 - Oversampling

---

[3] Modern filters do a pretty good job; still, we only have from 20 kHz up to 22.05 kHz to achieve about 100-120 db attenuation.

Unfortunately, while this pseudo-sampling *does* double the *data rate*, it provides no help with beat-frequency modulation—Figure 11.7 shows this type of oversampling applied to Fig. 11.5; so, how *do* we over-sample in a way that helps?[4]

If we transform the data shown in Fig. 11.5 we will get the spectrum of Fig. 11.8 (we're only taking the positive frequencies here, but we know there is a *companion negative frequency* component just above the Nyquist). If we now reconstruct the time domain signal we will get the waveform we started with; however, bearing in mind that the frequency domain is (a phasor representation of) the sinusoids that make up the time domain data, it's apparent we may reconstruct the sinusoids (i.e., the time domain signal) at twice the original sampling rate (i.e., calculate the data points at half the original data intervals)....[5] If we have obtained the real harmonic components in the transform, *and we haven't done something wrong again*, it shouldn't matter what sample rate we use to *re-create* the time domain data. The neatest part of this scheme is that we can obtain the desired oversampling by simply increasing the size of the frequency domain array (*packing* the upper locations with zeros). Fig. 11.9 shows the resulting spectrum of the *oversampled* data (here we've increased the sample rate by a factor of 16:1). The single harmonic of Fig. 11.8 lies near the bottom of *this* spectrum. Reconstructing the time domain waveform we get Fig. 11.10, and the intermodulation distortion is gone (without filtering).

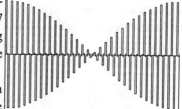

Fig. 11.7 - Oversampled Data

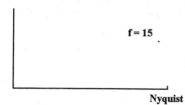

Fig. 11.8 - Transformed Data

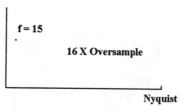

Fig. 11.9 - Oversampled Data

---

[4] We discuss how this is done in commercial CD players at the end of this chapter, but our first interest is Fourier transforms.

[5] We can reconstruct at 4× or 8× the original data rate too, of course.

Digital Audio

Program FFT11.01 is designed to generate this type of oversampling, and we will need to experiment with it a little (as will become apparent shortly). Initially, it appears we can solve the beat-frequency problem if we oversample by 16:1 (actually, 8:1 does the same thing);[6] but, unfortunately, this illustration is (once again) just a little *too* simple. The problem here is that, as with spectrum analyzers, we must deal with real world signals, and these are not so well behaved as the signals of Figs. 11.5 & 11.10.

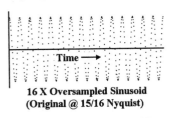

16 X Oversampled Sinusoid
(Original @ 15/16 Nyquist)

Fig. 11.10 - Reconstructed Data

11.3 THE MALEFICENT GIBBS

As we saw in Chapter 9, things get messy when frequencies are not exact integer multiples of the time domain fundamental, and while musical sounds *are* composed of *quasi-integer multiple* frequencies, as far as the Fourier transform is concerned, they're not integer *at all!* In Fig. 11.11 we show the now familiar spectrum of a *non-integer multiple* harmonic, and Fig. 11.12 shows the oversampled spectrum. Figure 11.13 shows what happens when we reconstruct this oversampled, non-orthogonal, fractional-frequency sinusoid... =(Oops!

Fig. 11.11 - 4.5 Hz Spectrum

The reconstructed fractional frequency (that has been oversampled) of Fig. 11.13 is, of course, incompatible with high fidelity sound—this will never work, but *why* does this fractional frequency component reconstruct with so much distortion? *[We know that, without oversampling, fractional frequencies reconstruct perfectly].*

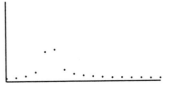

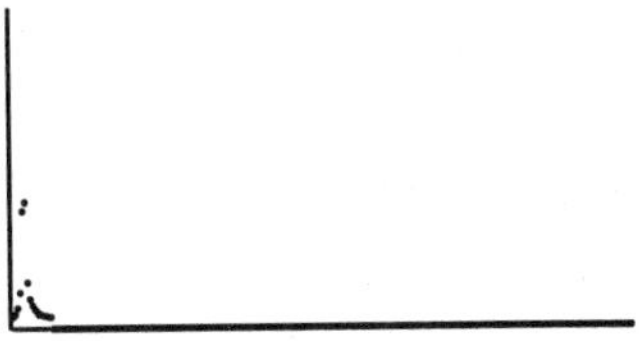

Fig. 11.12 - 16 X Oversampling

If you look closely at Figure 11.12 you will see that the spectrum looks very much like the *truncated* spectrums we created when we

---
[6] We use 16 X here to make a pretty picture. In practical systems this would not need to be more than 4 X (or maybe 8 X in a "top-of-the-line" system).

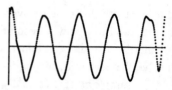

Fig. 11.13 - 4.5 Hz Oversampled

illustrated Gibbs phenomenon—the upper 15/16 of this spectrum is *packed with zeros*. If we had actually *sampled* a 4.5 Hz signal at the oversampled rate the spectrum wouldn't simply truncate at 1/16 of the available frequency space. Apparently the distortion we see in Fig. 11.13 is a close relative of Gibbs! Now, we have seen how we can alleviate this problem in previous chapters, and we will investigate that possibility shortly; however, we will never be very good at FFT applications unless we understand it's internal workings. We have a golden opportunity to look inside the Fourier transform here, so let's listen to ole' Yogi and take this *"fork in the road."*

11.4.1 ABOUT THE DFT

When we oversample as we have just done, we are creating data points *between* the original digitized data points, based only on the spectrum obtained from those original data points! This seemed like a good approach initially but now we have encountered problems, and we need to look a little closer at the underlying mechanisms.

When we studied impulse functions in Chapter 10 (*pff.* 112) we found that equal amplitude harmonic components summed to an impulse at the initial data point, but cancelled to zero at all following data points. If you review Table 10.2 (*p.* 113), it's obvious that, at points between the data points *shown*, these sinusoids will probably *not* cancel to zero! Table 10.2 can tell us much about what's going on here, but first let's explore this mechanism by considering a couple of theorems.

11.4.2 THE SHIFTING THEOREM

When we first discover the theorems of Fourier analysis they seem like magic; but, in truth, they expose simple relationships between the time and frequency domains. For example, *time domain shifting* (we used this in developing the FFT algorithm). This theorem says: *if we shift a time domain function (in time), its frequency domain function will experience a linear shift in phase proportional to the harmonic number.*

Let's use the impulse function to examine the mechanics of this

Digital Audio

relationship. The reason we obtain an impulse at the beginning of the time domain (Table 10.2) is that all the cosines have an argument of zero there. All other data points are zero because the harmonics shift *incrementally over $2N\pi$ radians* (because of their different frequencies). That is, going down any column of frequencies, the cosine functions are shifted by equal increments *over a number of full cycles*. Now consider the table below:

	T00	T01	T02	T03	T04	T05	T06	T07	T08
F00	+1.000	+1.00	+1.000	+1.000	+1.000	+1.000	+1.000	+1.000	+1.00
F01	+0.924	+1.00	+0.924	+0.707	+0.383	-0.000	-0.383	-0.707	-0.92
F02	+0.707	+1.00	+0.707	-0.000	-0.707	-1.000	-0.707	+0.000	+0.70
F03	+0.383	+1.00	+0.383	-0.707	-0.924	+0.000	+0.924	+0.707	-0.38
F04	+0.000	+1.00	-0.000	-1.000	+0.000	+1.000	-0.000	-1.000	+0.00
F05	-0.383	+1.00	-0.383	-0.707	+0.924	-0.000	-0.924	+0.707	+0.38
F06	-0.707	+1.00	-0.707	+0.000	+0.707	-1.000	+0.707	-0.000	-0.70
F07	-0.924	+1.00	-0.924	+0.707	-0.383	+0.000	+0.383	-0.707	+0.92
F08	-1.000	+1.00	-1.000	+1.000	-1.000	+1.000	-1.000	+1.000	-1.00
F09	-0.924	+1.00	-0.924	+0.707	-0.383	-0.000	+0.383	-0.707	+0.92
F10	-0.707	+1.00	-0.707	-0.000	+0.707	-1.000	+0.707	+0.000	-0.70
F11	-0.383	+1.00	-0.383	-0.707	+0.924	-0.000	-0.924	+0.707	+0.38
F12	+0.000	+1.00	+0.000	-1.000	-0.000	+1.000	+0.000	-1.000	-0.00
F13	+0.383	+1.00	+0.383	-0.707	-0.924	+0.000	+0.924	+0.707	-0.38
F14	+0.707	+1.00	+0.707	-0.000	-0.707	-1.000	-0.707	-0.000	+0.70
F15	+0.924	+1.00	+0.924	+0.707	+0.383	+0.000	-0.383	-0.707	-0.92
	0.000	16.00	0.000	0.000	0.000	0.000	0.000	0.000	0.00

TABLE 11.1 - Summation of Phase-Shifted Sinusoids

When we shift the impulse one data location to the right, the harmonics must all shift such that their *zero arguments* align in the second data position. Going across the rows we read the time-sampled harmonic components, so the top row (i.e., F00) is the *constant term* and the second row (i.e., F01) is the *fundamental* (again, only the first half of the time domain is shown). Note carefully that, for the fundamental, the "distance" between data points represents a phase increment of 22.5° (i.e., 360°/16 = 22.5°); however, the second harmonic has 45° increments, and the 3rd gets 67.5°, etc. To get the impulse in the second column, we must shift every harmonic one *position* to the right, so the following *must* occur:

1. The fundamental shifts by a phase of $2\pi/N$ where N is the number of data points in the sample interval (in Table 11.1, as we said, the phase shift is 22.5°).
2. The 2nd harmonic shifts by twice that amount (i.e., 45°).
3. The 3rd harmonic shifts by three times as much as the fundamental.
4. Etc., etc.

(Note that the first <u>column</u> is replaced with the data that would naturally occur at the phase shift given to each sinusoid.)

In general, each harmonic is shifted by $2\pi F/Q$ (where F is the harmonic number and Q is the number of data points). Furthermore, it's relatively obvious this relationship must hold between *any time-shifted* function and its frequency domain counterpart—it's completely general—it applies to any function.[7] It's also bilateral (as are all the theorems). That is, if we shift a data point in time the harmonics must shift linearly in phase; and, conversely, the equivalent phase shifts in the frequency domain must move the data points in time. The linear phase shift phenomenon is nothing but the result of shifting all of the different frequencies by the same time increment! No magic here—it's simply what happens.

11.4.3 THE ADDITION THEOREM

If the sinusoids of an impulse function sum to zero at the *second data point* (Table 10.2) then, if we add in the spectrum of another impulse which has been shifted one position to the right (Table 11.1 above) the summation of both spectrums *at this second data point* will simply be equal to the summation of the components of the 2nd spectrum (the summation of the 1st spectrum yielding zero as we know). But then, when we sum two spectrums, the value of every data point will be equal to the summation of the values of the sinusoids for both spectrums. In this specific case, summation of the two spectrums will yield an impulse in the first *and* second data positions (but still zero in all others). *Note: We have said a lot in this short paragraph, and it's important; so, if you need to go back and re-read it (especially that second sentence) it's okay.*

Now, since we know sinusoids of the same frequency can be added by simply adding their phasors, and we know that the DFT is nothing but phasors (i.e., the amplitude coefficients for each harmonic's complex components), then we can accomplish the summation of any two time domain functions by the summation of their Fourier components. From Tables 10.2 and 11.1, which show the explicit relationship of the frequency domain to the time domain, it should be clear this may be done

---

[7] Since any real function may be represented by a unique summation of sinusoids that are integer multiples of the fundamental, the phase shift between each sampled point will always be as described above (i.e., $2\pi F/N$). If we time-shift *any* function the sinusoids of its frequency domain representation *must* shift by the rules given above.

Digital Audio 129

for *any* two frequency domain representations. It should be clear that summing the DFT coefficients is the same as summing the sinusoids themselves—which is the same thing as summing the time domain functions. This obviously bilateral relationship is the *Addition theorem* (also known as the *Superposition theorem*), and there's certainly no magic here—only a time/frequency domain relationship.

11.4.4 THE DFT AS IMPULSES

So then, if we sum in the individual spectrums of N sequentially time-shifted impulse functions, we will certainly get the spectrum of the sum of N sequentially time-shifted impulses (if you're finding these statements more and more difficult it's because I've been advised to make this stuff sound more like a typical textbook). Consider this: every *function* that's composed of discrete data points may be considered to be the summation of N time-shifted impulses of amplitude A_n (where A_n is the amplitude of data point *n*). For a single impulse (i.e., a single data point), the amplitude of each harmonic component is equal to A_n/Q [where Q is the number of data points (and harmonics) in the array]. Our point is that every data point, when considered as a separate impulse, has the "same" transform (i.e., a flat spectrum of amplitude A_n/Q), except that each harmonic is shifted in phase to get the impulse into the right position.

Now, summing time domain signals is equivalent to summing their transforms, and since any digitized signal is nothing but a summation of time-displaced impulses, it's apparent that any signal we can digitize must always have a real, unique, frequency domain transform. It's simply the summation of all the transforms of the individual data points! That is, a spectrum of equal amplitude cosine waves unquestionably produces an impulse; and shifting the phases of an impulse spectrum will shift the impulse in time—and if we add sequentially shifted impulses we can reproduce any digitized data...and, if all this is true, then every digitized function *obviously* has a transform![8]

There's a specific reason we've gone to this trouble, and it hinges on the ultimate nature of this impulse function we have been considering here; so, without further ado, let's now consider an oversampled impulse function.

---

[8]It's apparent, then, that if stretching the spectrum of an impulse function duplicates the impulse, stretching the spectrum of *any* function will duplicate that function.

11.4.5 THE IMPULSE/SINC FUNCTION RELATIONSHIP

In FFT11.01 we've provided an *impulse function* generator. The impulse function, of course, yields a flat spectrum of unit amplitude cosines (Fig. 11.15). If we oversample this spectrum we get the result pictured in Fig 11.16. Reconstruct the time domain and, voilà—you have a *sinc function* (see Figure 11.17). *[Note: The DFT/FFT thinks all data is periodic, whether it is or not, and we're seeing the beginning of the next impulse which is the part of the impulse that occurred prior to the beginning of our data sample.]*

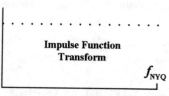

Fig. 11.15 - Impulse Xform

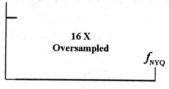

Fig. 11.16 - 16 X Oversampled

This is reminiscent of the result obtained when we oversampled a fractional frequency; so, let's take a closer look at *why* we get a sinc function instead of an impulse when we oversample. Let's return to that impulse function table—the one we've been using to investigate the internal mechanics of the DFT—and look at what happens when we generate data points *between* the original digitized points. We need only look at the leading edge (see Table 11.2), and we will consider points that are generated by 8× over-sampling. Note that the zero time and zero frequency still yield values of 1.00, but now only the 4th and 8th data points sum to 0.00. A careful perusal of this table will tell you why a sinc function appears at the oversampled data points. [We will accept that this is, in fact, a sinc function (i.e., sin(x)/x—this is derived in the text books).]

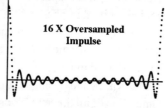

Fig. 11.17 - Reconstruction

So, when we *oversample* a signal, we're doing something very similar to when we *whack-off* the upper end of the spectrum and create the Gibbs phenomenon; however, we're doing more than just that. We are (literally) looking *between* the data points—at the continuous function we called previously "an incomplete summation of sinusoids." This is similar

Digital Audio

to where, in Chapter 10 (with fractional-frequencies), we looked between the pickets in the frequency domain, but with oversampling we're looking between the pickets in the time domain.[9]

	T00	T01	T02	T03	T04	T05	T06	T07	T08
F00	+1.000	+1.00	+1.000	+1.000	+1.000	+1.000	+1.000	+1.000	+1.00
F01	+1.000	+0.99	+0.981	+0.957	+0.924	+0.882	+0.832	+0.773	+0.70
F02	+1.000	+0.98	+0.924	+0.832	+0.707	+0.556	+0.383	+0.195	+0.00
F03	+1.000	+0.96	+0.832	+0.634	+0.383	+0.098	-0.195	-0.471	-0.70
F04	+1.000	+0.92	+0.707	+0.383	+0.000	-0.383	-0.707	-0.924	-1.00
F05	+1.000	+0.88	+0.556	+0.098	-0.383	-0.773	-0.981	-0.956	-0.70
F06	+1.000	+0.83	+0.383	-0.195	-0.707	-0.981	-0.924	-0.556	-0.00
F07	+1.000	+0.77	+0.195	-0.471	-0.924	-0.957	-0.556	+0.098	+0.70
F08	+1.000	+0.70	-0.000	-0.707	-1.000	-0.707	-0.000	+0.707	+1.00
F09	+1.000	+0.63	-0.195	-0.882	-0.934	-0.290	+0.556	+0.995	+0.70
F10	+1.000	+0.56	-0.383	-0.981	-0.707	+0.195	+0.924	+0.831	-0.00
F11	+1.000	+0.47	-0.556	-0.995	-0.383	+0.634	+0.981	+0.290	-0.70
F12	+1.000	+0.38	-0.707	-0.924	-0.000	+0.924	+0.707	-0.383	-1.00
F13	+1.000	+0.29	-0.832	-0.773	+0.383	+0.995	+0.195	-0.882	-0.70
F14	+1.000	+0.19	-0.924	-0.556	+0.707	+0.831	-0.383	-0.981	+0.00
F15	+1.000	+0.10	-0.981	-0.290	+0.924	+0.471	-0.831	-0.634	+0.70
	16.000	10.68	1.000	-2.871	0.000	2.496	1.000	-0.897	0.00

TABLE 11.2 - Summation of Over-Sampled Impulse

Now, in the above illustration, we have deliberately used a single data point so the sinc function phenomenon will be apparent; but, when there is more than one data point, each of them will have its own time domain sinc function. The oscillatory nature of all these sinc functions makes them tend to cancel—*except* at edges—at boundaries—similar to the way Gibbs phenomenon appears at discontinuities. We might, therefore, reasonably expect the problem to be greatly relieved by using greater array sizes.[10] While we would still expect to see the distortion near the boundaries of the data array we should see a greatly reduced effect throughout most of the reconstructed data. This provides us with an avenue to pursue the problem.

---

[9] This similarity between the time and frequency domain sinc functions is very fundamental. When we deal with a finite number of data points with a finite number of harmonic components (i.e., the DFT), a sinc function (which, as we know, exists as sidebands about the harmonic components simply because the data starts and stops) also appears about time domain data points when their *spectrums* start and stop. We have noted the similarity between the forward and inverse transforms—except for the negated sine component and a multiplied constant, they're identical. We shouldn't be too surprised, then, if a truncated spectrum produces a sinc function about each data point.

[10] As the array size increases the number of harmonics increases, and we might reasonably expect the distortion to shrink toward the edges just as the Gibbs phenomenon shrinks toward the edges as we add harmonic components.

11.5 SOLVING THE OVERSAMPLING PROBLEM

First of all, the advent of the DVD format makes solving this problem (as we have currently framed it) unnecessary,[11] but there are other reasons for pursuing this technology. For example, when loud-speakers are driven outside of their optimum frequency range the overall sound tends to be degraded. Speakers are sometimes connected using filter networks to attenuate frequencies outside the optimum band, but it's difficult to do this effectively. In our ultimate system we could separate frequencies via the FFT and drive the speakers with an optimum band of frequencies, but here oversampling would be unavoidable.

Now, *every* sinusoid in an audio signal must have *some* function's sidebands (since the data starts and stops); but, when we transform blocks of data we *introduce* sinc function sidebands. When we look between the original data points (i.e., we *oversample*) we will see these (except when they cancel each other). One way to minimize this phenomenon, we know, is to use a weighting function before taking the transform. As we saw with our spectrum analyzer, this produces tight little sidebands...which, hopefully, will not be too distorted if we oversample and reconstruct. There *is* one problem—when we use this technique what we transform is not the same thing as the original audio [i.e., we will transform (for example) a $\cos^2$ *weighted* version of the audio]. But this is no big problem—on reconstruction we can *de-weight* the reconstructed signal (i.e., *divide* by the weighting function). So, when we oversample this weighted function spectrum, the effect we encountered earlier should be greatly reduced. We find, in fact, with weighting and de-weighting, distortion *is* attenuated, but not completely *eliminated*. There

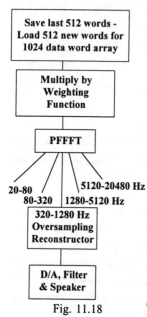

Fig. 11.18

---

[11] DVDs alleviate this problem by storing data at a rate of 96 kHz. Via a compressed data format they provide two hours of 5 channel, 24 bit words at this 96 kHz—truly high fidelity sound. We will discuss the benefits of using 24 bit words shortly, but we should see the eminent demise of the *CD format* in this.

Digital Audio

are still vestiges of the problem near the beginning and end of the reconstructed signal. So, how shall we solve this problem?!!

[If we, say, load 1024 data words into a buffer and oversample as we described above, on reconstruction we know we will get distortion at the *seams* of these buffers. If, however, we have really fast hardware, we can transform, oversample, reconstruct, and then *throw away the first and last quarters* of the reconstructed data.[12] Only the middle half would go to the D/A. We would then load the *second half* of the data to the *first half* of the data array, load the next 512 data points into the (last half) of the array, and repeat the process. If the transient distortion has decayed far enough after 256 data points, we get *practically* seamless output data.]

11.6 OVERSAMPLING VIA CONVOLUTION

Let's look at how oversampling is accomplished in practical CD players. The FFT is not used, but rather, an algorithm based on the convolution integral we discussed in the last chapter of *Understanding the FFT*. Since the data we end up with will be oversampled (i.e., we will have extra data points), we know a sinc function will exist about these data points. To *de-convolve* this oversampled data we essentially must convolve it with another sinc function. Program APS11.0 illustrates how this works:

```
' ****************************************************
' *        APS11.00   OVERSAMPLE ILLUSTRATION         *
' ****************************************************
'THIS PROGRAM GENERATES A GRAPHIC OF A SAMPLED SIGNAL NEAR THE
'NYQUIST.  THE SECOND PART OF THE ILLUSTRATION CONVOLVES
'THE OVERSAMPLED DATA WITH A SINC FUNCTION.
10 SCREEN 11: CLS : PM = 1: F1 = 0.9
PI = 3.141592653589793#: PI2 = 2 * PI: K1 = PI2/200
DIM C(128)
X0 = 20: Y0 = 20: XX = 620: YX = 320: YM = 220
LINE (X0, Y0)-(X0, YX)' DRAW Y AXIS
LINE (XX, YM)-(X0, YM)' DRAW X AXIS
20 FOR N = 0 TO 20.01 STEP .01  ' CREATE SAMPLED DATA PATTERN
X1 = N * F1 * PI: Y1 = COS(X1)   ' CALCULATE DATA POINT
LINE -(X0 + N * 600 / 20, YM - Y1 * 150)' DRAW SINE WAVE
22 IF CINT(N) = CINT(100*(N))/100 THEN GOSUB 60 ' SAMPLE DATA
NEXT N
INPUT A$  ' WAIT FOR USER PROMPT
'  ***** 16XOVERSAMPLE & CONVOLVE DATA WITH SINC FUNCTION *****
' FIRST EXPAND DATA ARRAY
30 FOR N = 20 TO 40: C(N) = C(N - 20): NEXT N ' REPEAT 1ST 20
31 FOR N = 40 TO 80: C(N) = C(N - 20): NEXT N ' REPEAT 1ST 40
```

---

[12]We may not need to throw so much away—this is just "brain-storming."

```
31 FOR N = 40 TO 80: C(N) = C(N - 20): NEXT N  ' REPEAT 1ST 40
32 FOR N = 20 TO 40  '*** START OVERSAMPLE - CONVOLUTION ***
34 FOR D = 0 TO .875 STEP .125
36 IF D = 0 THEN Y2 = C(N) ELSE GOSUB 62
38 FOR I = 1 TO 20
Y2 = Y2 + (C(N+I)*SIN(PI*(I-D)) + C(N-I)*SIN(PI*(I+D)))/ (PI*(I+D))
NEXT I
40 CIRCLE (X0 + ((N + D - 20)*600) / 20, YM - (PM*Y2*150)), 2, 14
NEXT D
NEXT N
INPUT A$: STOP
60 '* SUBROUTINES TO SAMPLE AND PLOT DATA (EVERY 20TH DATA POINT) *
LINE (X0 + N * 600 / 20, YM)-(X0 + N * 600 / 20, YM - Y1 * 150)
C(N) = Y1  ' SAVE SAMPLED DATA POINT
CIRCLE (X0 + N*600/20, YM - (PM*Y1*150)), 4, 15  ' DRAW DATA POINT
RETURN
62 Y2 = (C(N)*SIN(PI*D)/(PI*D)) + (C(N+1)*SIN(PI*(1-D))/(PI*(1-D)))
RETURN
```

Let's cover this quickly—lines 10 to 20 generate constants, etc. At line 20 we start generating (and drawing) a cosine wave, and at line 22 we test to see if the variable N is an integer. If so, we jump down to line 60 and *sample* the data (a circle drawn on the signal indicates the sample point). The program then halts and allows the user to examine the result.

At the user's prompt we oversample and convolve this data. Since this data extends in both directions, we extend the data array (lines 30 & 31) giving the program access to past and future data points. At line 32 we begin the oversample-convolution algorithm. We will generate four data points for every one that exists in the original array by stepping one quarter of the time interval between the original data points (line 34). Now, for the original data points themselves, we know the sinc functions will all be zero, so when $D = 0$ we need only transfer the data point to the output array—if $D \neq 0$ we are generating an oversampled data point and jump down to line 62 to approximate the initial value of this data point. We set this data point to a *sinc function* weighted average value of the two data points on either side. We then move to line 38 and complete the convolution by summing in the sinc function weighted values of the data points (going in both directions—as we explained in *Understanding the FFT*). The result of this routine is then plotted on the same graphic we generated before, and the results are as shown in Fig. 11.19.

This program does indeed essentially remove the beat frequency modulation, but it obviously fails to yield perfect results—and the reasons why are not hard to see. First of all we have only oversampled by 4:1, but more importantly we have only extended our convolving sinc function to

Digital Audio

plus and minus 20 data points. If you take the trouble to extend this algorithm to ± 30 data points you will find the approximation to the original sinusoid is considerably better. If, however, we suppose harmonics generated by the distortion in the reconstructed oversampled signal are integer multiples of the sinusoid itself, we realize this distortion will be well above the audible range.

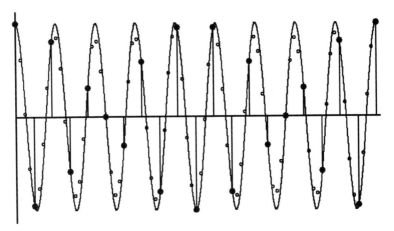

Figure 11.19 - Oversampling Via Convolution

Use of the Fourier transform in the above development has been only to analyze the nature of the problem; nonetheless, we recognize this convolution algorithm actually does the same thing as manipulating the frequency domain signal—in this case it's simply better to accomplish the convolution in the time domain. If a solution to an engineering problem exists, an infinite number of solutions exist. We can frequently discover how to build a better mousetrap by simply exploring this fundamental theorem of engineering.

PART II

AMPLITUDE RELATED PROBLEMS

We must talk a little about digitized signals, and the best way to do that is to review how a typical A/D converter digitizes a signal.[13] A simplified 3 bit "flash converter" is shown below:

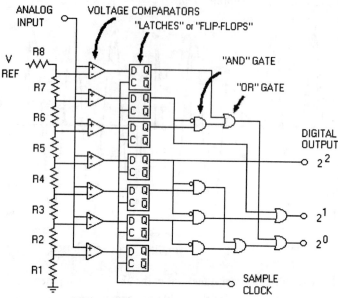

Figure 11.20 - Flash A/D Converter

There are eight identical resistors (R_1 through R_8) in the "divider string." If V_{REF} is equal to 8 volts (and circuit loading is negligible) the *voltage comparators* will have thresholds increasing incrementally in 1 volt steps (the triangle shaped functional blocks compare two voltages and output a "1" if the + side is greater—otherwise "0"). Now, during normal operation, the analog input voltage will vary between 0 and 8 volts. When the *comparison voltage* is exceeded, the comparator will output a "1." The clock pulses occur at a constant, precise rate, strobing the states of the comparators into a series of "latches" or "Flip-Flops" (i.e., bistable multivibrators). Note that when the input voltage exceeds the threshold of some comparator, turning it *on*, all comparators below it will also be turned *on*. The logic to the right of the latches encodes the state of the latches

---

[13] A/D stands for analog to digital. There are many different ways to convert analog to digital data but a *flash converter* is easy to understand.

Digital Audio

into a 3 "bit" binary number output.[14]

We muse now consider the A/D "transfer" curve (Fig. 11.21) of a typical A/D converter. Note that the output reads 000 until the input voltage exceeds 1.00 volt. It then reads 001 until the input exceeds 2.00

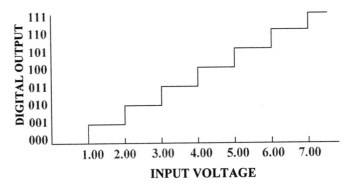

Figure 11.21 - A/D "Transfer" Curve

volts, etc., etc. This represents a form of *truncation error*, where the fractional portion of the input is truncated, resulting in the digital output always being lower than the input (except when the input *exactly equals* a transition voltage). If the input varies continuously the output will be, on average, 1/2 of a least significant bit (LSB) too low. When working with a 16 bit A/D converter, this error (which corrupts the 5th decimal place) may be negligible. If, on the other hand, we're working with an 8 bit converter, where the error corrupts the 3rd decimal place, it just might be a problem (in our 3 bit converter it's a serious problem).[15]

This problem can be alleviated by changing the threshold voltages. If the two *end resistors* in the voltage divider string of Fig. 11.20 are made 1/2 the value of all the others, and the reference voltage is reduced to 7 volts, the first threshold will be at 0.5 volts, the next at 1.5 volts, the 3rd

---

[14] It works like this: AND gates (flat backs) put out a *1* only if both inputs are *1* (i.e., one input is a *1* AND the other input is a *1*.) OR gates (concave backs) put out a *1* if either input is a *1* (i.e., one input is a *1* OR the other input is a *1*.) The little circles *negate* the input (i.e., turn *1* into *0* and *0* into *1*. Most often a *1* is 3 to 5 volts and a *0* is less than 0.5 volts.

[15] When we perform cumulative computations on data the truncation error may be significant in *any* digital system. The error accumulates at an average rate of 1/2 LSB per computation. Obviously, then, the right-most digit is compromised immediately, and ten successive computations may discredit the second decimal place. Even if your software returns 7 decimal digits, if you're doing a *lot* of computation, the last two digits may be inaccurate (if you need more accuracy you may want to use double precision arithmetic).

at 2.5 volts, etc., etc. The output will now be 001 for any input from 0.5 volts to 1.5 volts, 010 (binary 2) for inputs from 1.5 to 2.5, etc. The average values are now correct, but we haven't completely solved the problem; in fact, we have exposed a more fundamental error—the *digitization error*. When we convert continuous variables into discrete numbers, we encounter this digitization error. We can no longer "see" what the input is doing below the least significant digit. In binary systems this introduces an error of ± 1/2 of the least significant bit.[16]

So, then, how many bits do we need in our A/D converter? An 8 bit converter splits the input signal into 256 increments, and a 16 bit converter yields 65,536 different values. High fidelity *compact discs* use 16 bits for each sound channel and provides pretty good audio—*True Color* digital pictures use 8 bits for each of three primary colors. Our present example is digital audio, so let's look at this 16 bit digital word.

11.7 DYNAMIC RANGE

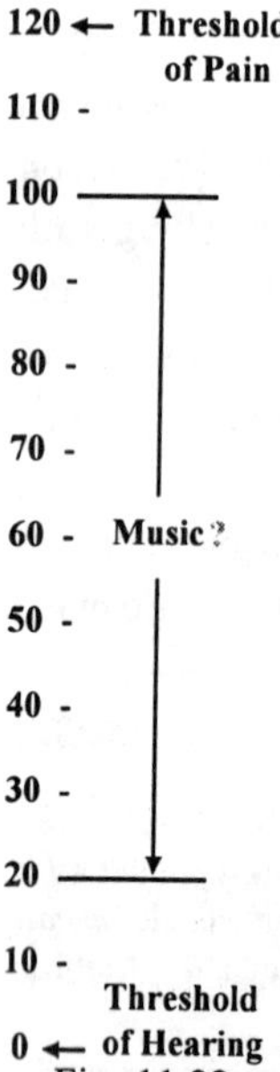

The dynamic range of the human ear is about 120 db (from the threshold of hearing to the threshold of pain). Now, playing music at the threshold of pain must surely imply masochistic tendencies—one would think we need not produce levels greater than about 120 db (above the threshold of hearing).

Unfortunately, the quasi-coherent nature of the harmonics in real music cause them to add in ways that make peak amplitudes 10-12 db higher than a simple analysis of dynamic range would indicate (Fig. 11.21). This extends the range to 130-135 db (?).

Neither do we listen to music at the threshold of hearing—if the sound is too low we turn the volume up. In the home, background noise usually makes threshold a moot point—in an automobile 40 db above threshold might very well be inaudible. This consideration doesn't reduce the dynamic range; but, rather, if we boost a quiet passage

Fig. 11.22

---

[16] Turning the average -1/2 LSB truncation error into a ±1/2 bit error doesn't completely eliminate the cumulative error. With a ±1/2 bit error the cumulative error then wanders off in a *random walk*. Fortunately, 5 decimal digits is more than adequate for most applications.

Digital Audio

by 20 db, noise (and distortion) are boosted by 20 db.[17]

Now, to optimize *signal to noise ratio*, recording studios could scan the digital master and make sure the highest peak amplitude is set to the maximum digital word size—but doing so would make the loudest passage in the *Moonlight Sonata* equal the loudest in the *1812*, and when we play a disc with more than one selection, most people would find this unacceptable. So, we lose 20-30 db here (so far as noise and distortion are concerned).[18] We consider this in greater detail later.

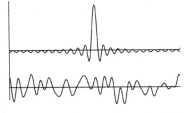

Figure 11.23 - Two signals composed of 32 harmonics of equal amplitude. Phases are randomized in the lower waveform.

So, we need about 130 db of *distortionless* dynamic range (i.e., practical considerations expand the range between 20 and 110 db considerably). [Note: "distortionless" doesn't mean *no distortion at all*! We need only guarantee that distortion is below the threshold of hearing.

11.8 DYNAMIC RANGE (LINEARITY DISTORTION)

CDs specify their dynamic range at 96 db—let's look at this specification. On a CD the signal is stored as 16 bit binary words (the data on the disc is stored in a reasonably elaborate encoding scheme but the *sound* data is still stored as 16 bit words). Now, we know that $2^{16} = 65{,}536$, so the largest *peak to peak* sinusoid that can be stored with this word size must fit between zero and 65,535 (effectively the most significant bit is a sign bit and the sinusoids vary between ±32,767 but let's not get bogged-down in details). The smallest possible *peak to peak* signal would be a change of 1 bit, yielding a dynamic range of $20 \, \text{Log}_{10}(65{,}535) = 96.3$ db. Okay, it's apparent where they get the 96 db, but let's not kid ourselves—a signal switching between zero and one is a square wave (see Fig. 11.24). Surely this is a very low sound level (about 20 db above the threshold of hearing if the max. amplitude = 120 db, but if you reach over

---

[17] Again, the real situation is complicated—threshold of hearing is frequency dependent, loud sounds "mask" soft ones, etc. (see, for example, Chapter 2 of Ballou's *Handbook of Sound Engineering*, Howard W. Sams).

[18] Things could be greatly improved with an embedded "loudness" (i.e., gain) word which could be updated every 0.1 seconds or so.

and turn up the volume on your amplifier, this will *not* sound musical (if your voice sounds like this don't bother sending your resume to the "Met"). Now, *we're* not so much concerned with such a simple-minded determination of dynamic range, but with *distortionless* dynamic range! Let's look at this just a little bit closer.

Let's look at how digitization error (see Fig. 11.21) affects digital sound. Not only do we acquire an error from the finite increment of the *digital steps* but we digitize at precise *times*, and the signal's crossing a threshold doesn't guarantee the A/D converter will record that fact. For example, by the time we perform the first A/D conversion in Fig. 11.24, the signal has exceeded two digitization increments, and the 2nd converted word includes *three* increments (the data, as noted earlier, is always digitized at *less* than the signal). This error, then, doesn't simply fail to capture the exact amplitude of the signal, *but actually records an incorrect form for the signal!* This is, of course, a non-linear distortion; but, if it doesn't repeat every cycle (in general it won't), it becomes quantization *noise*. As the quantization increment becomes smaller, and the sampling rate becomes faster, this form of distortion/noise rapidly decreases (otherwise CDs would be neither technically nor commercially viable). But how much distortion *do* we encounter? While this process isn't easy to analyze we have in our possession a powerful analytical tool—an FFT spectrum analyzer. We can model the whole A/D/D/A chain and then analyze the signal for harmonics (in the output) that shouldn't be there. (*This is too good to pass-up.*)

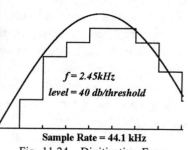

Fig. 11.24 - Digitization Error

11.9 MODELING THE D/A MECHANISM

We will put together a fairly simple model of the D/A output to illustrate this technique, but if you're really building digital sound systems you might want to consider this in greater detail. The D/A converter (for 16 bits) has maximum *peak* amplitudes of 2^{15} (i.e., ±32,768), so if we multiply a unit amplitude sine wave by 32,768, and take the *integer* portion

Digital Audio

of that sinusoid, we will pretty well duplicate the action of an A/D converter. We can then attenuate the sine wave before putting it into this A/D converter to simulate the digital signals. We will do all this with the following routine (see Appendix 11.3—FFT11-03):

```
400 REM GENERATE FUNCTION
402 FOR I = 0 TO Q-1:C(I) = 0: S(I) = 0: NEXT I ' CLEAR ARRAY
404 MAMP = 2 ^ 15: GAIN = 1 / MAMP: Y = 0: ISAMP = 0 ' SET-UP DIGITIZER
410 FOR I = 0 TO QDT ' GENERATE FUNCTION
414 Y = GAIN * INT(AMP * MAMP * SIN(F9 * K1 * I))
420 S(I) = Y
430 NEXT I
432 IF WTFLG = 2 THEN 450 ' USE WEIGHTING FUNCTION?
440 RETURN
```

In line 404 we set up a maximum possible amplitude of 2^{15} (MAMP) and a GAIN constant of 1/MAMP. The gain constant is used (in line 414) to convert the maximum signal back to 1.0, keeping our maximum signal at zero db for easy reference when reading the distortion.[19] In line 410 we set up a loop to generate the signal to be analyzed. We generate this *digitized* data at line 414 as we discussed above. The variable AMP must range from zero to a maximum of 1.0, and acts to attenuate the signal input so that we can see the effect of distortion for variable amplitude signals. In line 420 we place the value Y in the S(I) array and at line 430 jump back to generate the next data point. At line 432 we invoke the weighting function routine if it has been selected (just as in the standard spectrum analyzer of Chapter 9).

The *Generate Function* routine above is controlled by the *Generate Function & Analyze* routine shown below. In lines 602-604 we specify the frequency of the sinusoid to be analyzed. At line 612 we jump down to the *Generate Function* routine to simulate the D/A process.

```
'        *********************************
'        *   GENERATE FUNCTION & ANALYZE  *
'        *********************************
600 CLS : PRINT : PRINT
602 INPUT "SPECIFY AMPLITUDE, FREQUENCY (E.G., .8, 16)"; AMP, F8
604 F9 = F8 * FRACF ' CONVERT TO FRACTIONAL ARRAY DOMAIN SIZE
606 PRINT "PREPARING DATA INPUT - PLEASE WAIT!"
612 GOSUB 400 ' GENERATE SINUSOID
614 GOSUB 100 ' ANALYZE SIGNAL
```

---

[19] As we discussed earlier (Section 11.7 and Fig. 11.22), the maximum amplitude will generally be less than -10 db (and in many cases much lower), making the ratio of the signal to distortion/noise about 10 db less (*i.e., worse*).

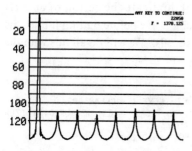

Fig. 11.25 - Synchronized Signal

We can now put this digital output simulation into our spectrum analyzer (we need only add it to the spectrum analyzer of the previous chapter) to see what distortion is generated. We find, as should have been expected, that non-linear distortion shows up as harmonics that were not in the original signal (*intermodulation distortion*, on the other hand, creates side band components). The *sum-total* of all these distortion harmonics must be below audible levels for a perfect system. You will find the worst distortions occur when the signal is an integer sub-multiple of the sampling rate (Fig. 11.25). This happens because the digitization non-linearity *locks in* and is identical for each cycle (as we explained on *p* 140); however, if the digitization error is not identical in sequential cycles, it shows up as digitization noise. Fortunately, this *lock-in* phenomenon is rare.

You will want to see what happens over both the frequency range and the dynamic range. The routine given in Appendix 11.3 will step through the chromatic scale, allowing amplitude to be changed so you may observe signal to noise and distortion ratios for various system conditions—it's an interesting exercise. You will need to know what happens if you reduce the size of the maximum digital word from 2^{15} to 2^{14} (or smaller). What happens when you *increase* the size of the digital word? [Oh, yes, and what happens as you change the resolution, etc., of the spectrum analyzer?]

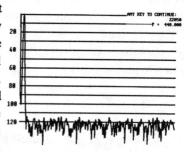

Fig. 11.26 - Non-Synchronized Signal

NOTE: *The coming of DVDs has changed things considerably. DVD audio addresses my major complaints concerning CD audio (i.e., primarily sampling rate and digital resolution, but it also expands on the technique of stereo sound). The DVD uses 24 bit words (16,777,216 values) for each channel, at a data rate of 96 kHz (or greater), which pretty well takes care of the problems discussed above.*

Digital Audio

11.10 RECORDING HIGH FIDELITY SOUND

The acoustic problems of recording sound today are no different than they were fifty years ago. If you have tried to record a performance you probably know that extraneous sounds can utterly destroy the recording (e.g., crinkling of paper, someone coughing, etc.). Technically, the way to capture a *high fidelity* recording is to lock the musicians in an *anechoic chamber*, but this approach greatly influences the musicians, and may change the final result dramatically). A more practical solution is to place a microphone just-off-the-nose of each musician, but this becomes cumbersome when recording a 100 piece orchestra. These acoustic problems, that plague the art and technology of recording sound, are much too far from our subject—we mention them here only to emphasize their impact on the following comments.

11.11 A GLANCE AT THE FUTURE

In the 21st Century we will surely see miraculous developments in high fidelity sound. Let's consider what might be forthcoming. A piano can be digitized and analyzed into perhaps a dozen or so harmonics (fewer

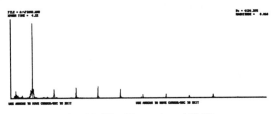

Fig. 11-27 - Piano A = 440 Hz.

in the higher register). To analyze this sound we need to digitize a single note at about 80 kHz for perhaps 20-30 seconds; however, once you analyze a note, it can be stored in about 200 data words (100 sine and cosine components). Now, the tone of a good piano should vary little between adjacent notes, so you would need to store only every tenth or twelfth key on the keyboard (the intervening keys would be reproduced by shifting the frequencies up and down by $\sqrt[12]{2}$ for each half tone. We might need a frequency domain weighting function (i.e., a filter) to represent the piano "box," but this would be a single constant. There are 88 keys on a piano keyboard which could be stored as 11 x 100 x 2 = 2200 data words. We could store a *high fidelity* symphony orchestra in perhaps 256 k-words (i.e., a mega-byte). We could then type-in the score to Beethoven's 9th (excluding the choral parts, of course) and hear a performance without all the imperfection accompanying the recording process. We could even glue an accelerometer to a baton and conduct the orchestra as we know it *should* be conducted (eat your heart out Arturo). Theater groups do *Oklahoma!* with little more than a piano—they could *use* a good synthetic orchestra.[20]

Along the same line of reasoning, we could put the cast of *La Bohème* in a carefully engineered acoustic room (using hearing-aid type earphones to hear the synthesized orchestra?), and record the voices of Rodolfo, Mimi, et. al. The synthesized orchestra, of course, could be added when the master is made; or, on playback, we could replay the voices, and synthesize a *perfect* orchestra at that time. This potential for synthesizing perfect orchestras eliminates the myriad problems associated with recording live performances.

Even more exciting, perhaps, is the possibility of enhancing very early recordings of famous artists. There are two approaches to the general problem of enhancement—one is to try to remove the noise and distortion captured in the original recording—the other is to replace each recorded *element*[21] with a *perfect element*. In the latter approach we must know *a priori* what the element *should* look like. We know what a piano (or trumpet, etc.) sounds like, so the problem reduces to *optimizing* the replacement element by minimizing the *error* between original and replacement according to some criterion. It would not be too difficult to measure the dynamics of a Rachmaninoff recording and synthesize his performance with a perfect piano—so too with the classic cornet solos of Bix Biderbeck.

In this new century we will also employ such techniques as separating the spectrum into perfect parcels and driving each speaker in the sound reproduction system with *only* its optimum frequency band—perhaps even measure the distortion of the speaker in this range and, prior to sending recorded sounds to a speaker, *compensate* for the distortion.

Surely we can look forward to virtually perfect (i.e., inaudible distortion) sound systems in the next 100 years.

---

[20] A simple program to synthesize a piano is given in the next chapter.

[21] Use of the term "element" will become clearer in the next chapter.

CHAPTER XII

CHANGING THE PLAYBACK RATE & SOUND SYNTHESIS

In Chapter 11 we considered the problem of changing the sampling rate while keeping the data unchanged (i.e., oversampling); but in this chapter we consider a much more difficult variation on that theme—the problem of slowing or speeding the data playback without changing the "pitch." Our investigation will bring us into contact with some interesting audio applications.

12.1 SIMILARITY

We may, by slowing down a record, turn a soprano into a bassoprofundo. As we know, in FFT jargon, this is the *Similarity theorem*—if we expand a function in time the frequency domain shrinks. This relationship is so fundamental and obvious it's hard to imagine how a recording could be slowed *without* turning sopranos into bassos, but our bout with *oversampling* hints at a crack in the armor—we added data points, but increased the *data rate*, yielding a *null result*...only the data sample *rate* was increased. Apparently we *can* manipulate the frequency domain data to achieve a desired result; but, compared to our present objective, that was child's play. Our problem here is considerably more formidable—we want to slow the record *without* turning sopranos into bassos—*we want to directly violate the Similarity relationship!* Now, *this* is almost sacrilege...sort of like breaking the laws of physics. (For those who believe man wasn't meant to have such knowledge, I'm writing this chapter to the ladies only—man reads it at his *own* risk.)[1]

12.2 THE FREQUENCY SHIFTING THEOREM[2]

We are well familiar with the *time shifting* theorem—if we *shift* a function in time, all of the harmonics shift in phase, proportional to the time shift *and* the harmonic number (see Section 11.4.2). There's a similar theorem about shifting in the frequency domain, of course, but before we discuss it we need to talk about the mechanics of shifting frequency components up and down.

---

[1] You never hear anyone say "Woman wasn't meant to have such knowledge!"(?)

[2] If changing the playback-rate shifts the frequency, it's more or less obvious we will need this—it's just not apparent *how* we can use it to break the law.

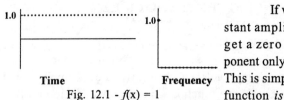

Fig. 12.1 - $f(x) = 1$

If we transform a constant amplitude function we get a zero frequency component only (Figure 12.1 left). This is simply what a constant function *is* (*forget* about the sinc function for now—we're in deep swampwater here!).

Okay, suppose we *shift* this zero frequency component up one notch in the frequency domain—up to the *first* harmonic position. We know this spectrum will reconstruct as a single sinusoid (Fig. 12.2). We will turn a constant amplitude into a sinusoid by shifting its transform (from zero) up one harmonic number. If we shift *this* harmonic up *another* notch, we will turn a single cycle of sinusoid into two cycles, etc., etc.

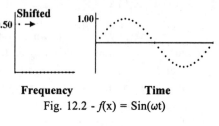

Fig. 12.2 - $f(x) = \sin(\omega t)$

Keeping this in mind let's now consider the *frequency shifting theorem*. What happens if we shift the spectrum of a more complicated function (e.g., a triangle wave)? *[Note: We have been using the PFFFT extensively in the last few chapters, but it will be best to revert to the conventional FFT for this illustration—we need the negative frequency components here.]*

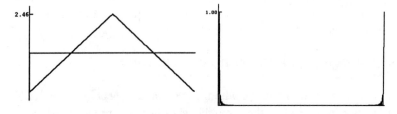

Fig. 12.3 - Triangle Wave

It was the negative sidebands (extending downward from the pulse train fundamental) that caused the beat-frequency modulation we struggled with in the last chapter. Clearly, these negative frequencies really exist, so if we frequency-shift the harmonics of a sampled data spectrum (i.e., any transform), we must also remember to include the negative frequencies.

Playback Rate

So, then, if we shift the spectrum of a triangle wave upward, we must fill the harmonic positions that have been vacated by the shift with the negative frequency components that lie just below zero (Fig 12.4). In practice we do this by shifting the negative frequencies *end-around* into the zero frequency position. This shifting mechanism creates a set of sidebands about a center frequency—a frequency equal to the frequency *shift*. This result is, of course, an *amplitude modulated* (suppressed) carrier. It's exactly as if we multiplied a sinusoid (carrier) with the original function. Note that the modulation of Fig. 12.5 is identical to the triangle wave of Fig. 12.3. Here we have increased the frequency without speeding-up the function! More to the point, we may shift the spectrum of Fig. 12.4 and only change the frequency—not the duration. This is not *quite* what we wanted, but it sure looks promising! It shows that the pitch and speed of playback are not eternally locked together—that while *similarity* surely describes a fundamental relationship, other relationships *are* possible. We will return to this line of investigation shortly, but for now let's look at the nature of the beast we have to tame.[3]

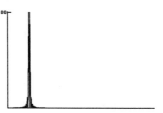

Fig. 12.4 - Shifted Xform

Fig. 12.5 - Frequency-Shifted Triangle

---

[3] Time domain theorems have frequency domain equivalents which, because of the similarity of forward and inverse transforms, are always similar. The similarity between the time and frequency shifting theorems may not be apparent as we have presented them; however, mathematically the time shifting theorem is written:

$$\text{Xform}\{f(t-t_1)\} = e^{-i\omega\, t_1}\, F(f) \qquad ------------ \quad (12.1)$$

and the frequency domain shifting theorem is written:

$$\text{Inverse Xform }\{F(f-f_1)\} = e^{-i\omega\, f_1}\, f(t) \qquad ------------ \quad (12.2)$$

In Section 11.4.2 we investigated time domain shifting sufficiently to know that Eqn. (12.1) says a time shift *is* a phase shift of the frequency components. Keeping in mind that the time domain function is just the summation of the harmonics, it becomes apparent that Eqn. (12.2) says a frequency domain shift produces a frequency shift of the time domain function. Note that Eqn. (12.2) multiplies the time domain function by a sinusoid—this is modulation—the equivalent of side bands in the frequency domain. That, of course, is exactly what we have produced in Figs. 12.3, 12.4 and 12.5 above.

Chapter 12

12.3 DISSECTING THE PARTS OF SPEECH

We're working with audio, so let's actually *look* at something *audible*. The time domain signal shown below is the audible function: "This is a test." At this scale it's hard to tell much about the individual sounds but there *is* something that's pretty obvious—this has the appearance of an *amplitude modulated* function. We note that the first three words are *slurred* together and, taken together, are only slightly longer than the single last word "test." We are severely limited in what we can do here—

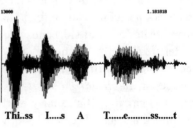

Fig. 12.7 - "This is a test."

we will not be able to handle whole phrases, but we *can* handle individual words (as *elements*). Let's take a closer look at "this."

Fig. 12.8 - "This"

In Figure 12.8 we see there is a distinct separation of this word into two parts—the "sss" sound is distinctly separate from the rest of the word. It's of much higher frequency and, as we see in Fig. 12.7, it's this "sss" sound that *bridges* the words in the phrase "thisisa." The "sss" sound is further expanded in Fig. 12.9.

Figure 12.9 covers a 10 ms. interval so the fundamental frequency appears to be around 4.5 kHz. With a sampling rate of 11 kHz (Nyquist = 5.5 kHz) we're marginal here. If we look at the first half of "this" word under the

Fig. 12.9 - The "sss" Sound

same microscope, we find *its* dominant frequency is about 400 Hz.

Okay, let's now look at the spectrum of this word. In Fig. 12.10 we see there are four *major* frequency components (ignoring the very lowest component). Each frequency *tick* on the x axis is 550 Hz. so the dominant component is around 400 Hz. Note that at about 4.5 kHz, we have something that looks like wide band noise. This, of course, is the "sss" sound shown in Fig 12.9. Finally, you will note the major harmonics

Playback Rate 149

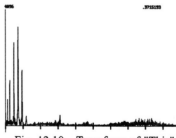

Fig. 12.10 - Transform of "This"

are *approximately* integer multiples of the lowest major harmonic frequency. These are produced by the voice-box, and must, of course, vibrate according to the laws of physics. It's unlikely they would not be integer multiples (approximately).

We may separate-out the major frequency domain components of this word and inverse transform them individually (see Figure 12.11). Note they all reconstruct faithfully with respect to position—even the "sss" sound in the bottom trace. These major components represent *sound elements* that make up spoken words (and most other common sounds too). Note that the *amplitude modulation* on these components is very low frequency—a couple of Hertz at most. We can *recombine* these sound elements, of course, and even though we eliminate everything else from the spectrum, we will obtain a reasonably good reproduction of the original function "this." If you do this and then play it back on your computer, it will actually sound like "This." (This particular reconstruction yields a slightly deadened sound, but a clever engineer/programmer could probably figure out why and fix that—surely these sound *elements* contain the complete sound.)

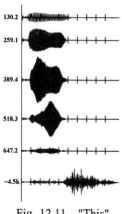

Fig. 12.11 - "This"

We should note that these speech components carry their own weighting functions, generally simplifying our task. This built-in weighting is part of the spoken word but, for our analysis, it means what we see is what we really want to see. Coincidentally, this built-in weighting function sort or *validates* our earlier decision to ignore the effects of the ever-present sinc function. Before we continue our investigation of variable playback rate let's digress to consider a very interesting problem that remains to be completely solved. Let's talk about *how* we might recognize speech.[4]

---

[4] I have no idea how the great gurus of *Speech Recognition* actually do this—this is only a "straw man" we can beat-up while we point out a few things. Our *primary* objective here is to develop the notion of *sound elements* as modulated sinusoids whose sidebands give a distinctive pattern.

12.4 SPEECH RECOGNITION

We might create a dictionary of *syllables* corresponding to the *sound elements* of the preceding section, then break each digitized word into its sound elements and *pattern match* (i.e., recognize) syllables (and/or phonemes). We would then construct (i.e., recognize) words from syllables, and probably phrases from the words. It may be, however, that pattern matching at the syllable level isn't sufficient—we may have to look at a lower level—perhaps try to recognize vowel sounds. If, for example, we could detect a "th" sound at the beginning, "sss" at the end, and a short ĭ sound for the middle vowel, we have "this." If the vowel is a short ŭ we have "thus." Now, these low level *elements* all sound pretty Neanderthal; still, these components *can* be recombined to get the original words, so the vowel sounds, *t* sounds, *p* sounds, etc., *must* be there somewhere.

Let's look at what we know so far:

1. The mid-range frequency (i.e., 300-900 Hz) major harmonics are the things that, together, give an individual voice its "identity." They generally coincide with the vowel sounds or "phonemes," and may be used as a "locator" for vowel sounds in words.

2. The "sss" sound is a wide band noise *hiss* which frequently blends words into *run-on* phrases (e.g., "thisisa").

3. *T*, *P*, etc., are distinctive elements which frequently identify beginnings/endings of words and syllables (e.g., *Test* or *Perpetuate*).

We can surely discover many such rules by studying phrases as we have done above, but there are a great many other problems in speech recognition. To illustrate we repeat the example of the previous section but spoken by a different person. Immediately we notice the *sss* sound is missing. The voice depicted in Fig. 12.12 was digitized at a rate of 11,025 samples/sec (Nyquist of 5,512.5 Hz). Now, AM radio bandwidth is only 5 kHz, and we would think *that* should be plenty for what we're doing; however, in this particular voice, the *sss* sound peaks at almost 9 kHz. We must sample at about 20 kHz (Nyquist of 10 kHz) and, indeed, the same voice, sampled at 22,050 samples/sec, recovers our *sss* sound. If we hope to do decent speech recognition, then, we will need high fidelity sound.

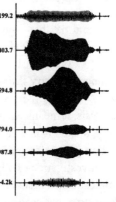

Fig. 12.12 - "This"

Playback Rate

So, what about recognizing speech by pattern matching these sound elements? Figs. 12.11 and 12.12 are close, but they're far from a perfect match. Nonetheless, when we combine these components, the words *are* reproduced—the *identity* must be in there.

Okay, this is a world-class puzzle. If you have a multi-media computer at your disposal, you have all the laboratory you need to do the R&D. You can record voice using any *sound board* and the program in Appendix 12.1 can be used to analyze and butcher the *.WAV files. *[If you decide to tackle any part of this problem you should look this program over carefully just so you know exactly what each function does—you might want to write your own version.]*

In any case, if you intend to use the Fourier transform to work on audio, you will have to butcher the files as described here. Here are a few hints that might be helpful: You can separate the words in run-on phrases by filtering off the spectrum above 2 kHz (i.e., by eliminating the "sss" sounds) and then reconstructing the phrase. You will need to add the "sss" back after separating the words, of course, and test to see which of the words it belongs to (e.g., "this is a sea shell").

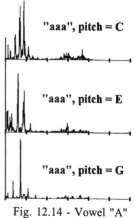

Fig. 12.14 - Vowel "A"

Recognizing vowel sounds is a particularly intriguing problem. You might want to record prolonged vowel sounds (i.e., "aaa", "eee", "iii", etc.) and investigate both their spectrums and time domain waves. You might need to say these at different musical pitches to find the tone invariant characteristic of each vowel.

Surely this is a world class problem, and its solution is, apparently, not out of our reach...if we have access to nothing more than good desk top computer....

12.5 SPEEDING UP THE PLAYBACK

So, what *was* our intention when we went out into that swamp? Well, what we *want* is to get the opera over in less time so we can go to bed. We can do this by playing the 33⅓ record at 45 RPM; but, that basso will turn into a soprano (i.e., the frequency domain will be expanded as the time domain shrinks.) Right—but why couldn't we just use the frequency shifting theorem to shift him back down to bass again? In fact, we can, but it's not completely straightforward. When we speed up the record, every frequency will be increased *proportionally* to the increase in playback rate (i.e., if we double the speed of the record, every frequency that was recorded will be *multiplied* by 2), but *frequency shifting* shifts all the frequencies by a *constant increment*. That is, the result of *shifting* is:

$$f_{er} = f_i + A \qquad \text{not} \qquad f_{er} = A f_i$$

To accomplish the result we desire we will have to recognize the distinction between these two, and use both mechanisms...and that is why we had to develop the notion of sound elements (as we shall see).

We have already considered a form of proportional shifting—the *frequency stretching theorem*. When we *stretch* the spectrum (by placing

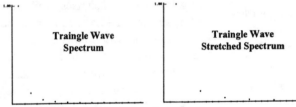

Fig. 12.15 - Triangle Wave Spectrums

zeros between every frequency component—see Figs. 12.15 above) we *duplicate* the data on reconstruction—we get a second copy (in the same time domain). If we reconstruct this stretched spectrum we will get *two* cycles of that triangle wave (if we had taken the transform of two cycles we would have gotten this *stretched* spectrum in the first place).

But suppose we start with the modulated carrier spectrum of the triangle wave we discussed back in Section 12.2 (see below—consider it

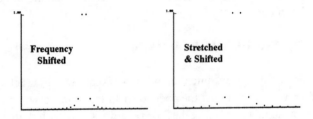

Fig. 12.16 - Stretched and Shifted

Playback Rate

to be one of our *sound elements*). Using what we have discussed, how can we speed up this triangle wave without changing the frequency? Well, if we had shifted the stretched waveform of Fig 12.15 by the same amount we shifted the original function (i.e., to the same center frequency—see Fig 12.16 above), then when we reconstruct this spectrum we will surely get the waveform shown on the right hand side of Fig. 12.17. Consider this carefully for in this case all we have done is stretch the modulation sidebands—the *carrier* frequency is unchanged. We get twice as many cycles in the same time interval but the frequency is the same...that sure *looks* like a light at the end of the tunnel.

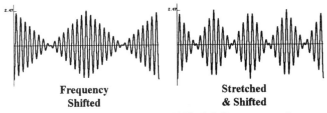

Frequency Shifted   Stretched & Shifted

Fig. 12.17 - Stretched & Shifted & Reconstructed

Okay, so, how do we use this trick to speed up the playback and keep our basso basso? Well, we have seen how we can decompose spoken words into *elements* based on their frequency domain function, and these components can then be recombined to yield the original words (see Fig. 12.11). These *major components* are simple enough that they are not significantly different from the triangle wave of Fig. 12.17. If, then, we transform any one of the components of Fig. 12.11, stretch the *sidebands* about the component's dominant frequency (there's no need to shift it down to zero, stretch it, and shift it back up again) then, when we reconstruct, these components will occur twice in the same time period (we will just throw away the back half). If we

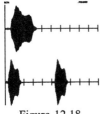

Figure 12.18

perform this operation for all major components and then recombine them (from *superposition* we can do *that* in either the frequency domain *or* time domain), we will have our basso singing the opera twice as fast, and he will still be a basso. *[We can go to sleep in two hours instead of...one?]*

But this raises another problem—if they gallop through the opera at *twice* the tempo that's too much (*Figaro* might be interesting though). What we really need is something maybe 12.5% faster (*that could still shave 30 minutes*). The problem is that, in a conventional DFT, we are limited by the discrete nature of the data. Trying to move between discrete data points presents problems (as we all know by now). We are fortunate in that we know about fractional frequency transforms, which allows a certain amount of maneuvering.

12.6 THE 12.5% SOLUTION

Suppose we use a 1/16 fractional frequency configuration—we know (from Chapter 9) this can be done by packing the last 15/16 of the array with zeros. We also know that every 16th frequency component will yield the same data as a conventional FFT, so it's apparent we could extract every 16th frequency component, put them into a smaller array, perform a *conventional inverse transform* and get the same data we started with (except for a 16:1 amplitude scaling factor). *If this is not obvious, go back and review Chapter 9!*

Okay, but we could also extract every *8th* harmonic, put these through a conventional inverse FFT, and we would then get the ½ fractional frequency configuration (i.e., data in the front *half* of the time domain and zeros in the back). In fact, that is the configuration we actually need—it will give us room to both stretch and contract the waveform; but, before we start transferring harmonics over to this *conventional half-frequency inverse FFT*, we need to look at how we're going to speed things up (and slow them down) in this scheme. We will keep the center frequencies of each *sound element* constant (just as we described in the preceding section), and we will expand (or contract) the *side bands* about each of these major components...and here's how we're going to do that in 12.5% increments: While we are still in the 1/16 fractional frequency array, we must pick out all of the major sound elements of the transformed voice. We then *extract* these major components, taking care to preserve some 10-20 full sidebands about each major component (i.e., that's 10 x 16 fractional frequency components). So, when we transfer over *every 8th fractional frequency component*, we will transfer the center frequency of every major component faithfully; however, instead of simply transferring every 8th sideband component, we might transfer every 7th sideband (counting plus and minus from the center frequency)...or, perhaps, every 9th, or.... We will select every 8th ± nth

Playback Rate

sideband depending on how much we want to speed up or slow down the playback. We will then be stretching the sidebands in 1/8th harmonic steps, and the playback will *compress* in 12.5% steps. Unfortunately, we are up against the limitations of QuickBasic™, and (when we consider people like Harry Nyquist and Edsel Murphy) we can't quite swing a 1/16th transform; but, we can still demonstrate the basic concept (see Appendix 12.1). Fig. 12.19 is the 3rd major component of the word "This" (from Fig. 12.11), and it has been made large enough (but just barely) to see the frequency remains constant.

To produce a practical system, of course, we must address the problems we mentioned under speech recognition (e.g., how to determine major components, how many sidebands to include, how and where to sum components, how to separate words, etc., etc., etc.) It might be desirable to handle all these problems prior to producing the CD and record the album (or more probably the audio book) in the transform domain. We could then reconstruct the audio at playback time, with the playback speed under control of the listener.

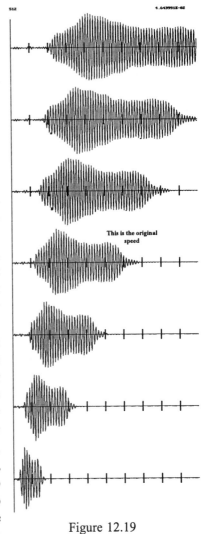

Figure 12.19

12.7 MUSIC SYNTHESIS

Music synthesis is the *flip-side* of changing playback rate in that here we must change the frequency of a given sound. We start with a model of the instrument to be synthesized and select the frequency to give the proper note. Let's look at the problems involved in synthesis.

We know that a piano, or trumpet, or etc., has harmonics in the sounds they make that give them their individual identities. Furthermore, we know these harmonics must be integer multiples of the fundamental tone (approximately) if the sound is to be musical. We already know that these harmonics are not perfect integer multiples, which contributes to the different sounds of different instruments; however, if we examine these "harmonics" closely, we find that they are not simple sinusoids—they are *sound elements* similar to those we discovered in the human voice. They have a microstructure which also strongly influences sound identity.

This, then, exposes the fundamental problem faced in music synthesis—we must preserve the *quasi-integer multiple* relationship of the harmonic elements *without changing the microstructure of the harmonic element itself.* So, when we synthesize notes, the center frequencies of the *harmonic elements* will be multiplied by the desired frequency step [i.e., $(\sqrt[12]{2})^N$], but the sideband components of the harmonic's microstructure are obviously *not* multiplied by this factor. If they were, we would be back to the speeding record syndrome! The sound *element* must be *placed intact* into the *position* determined by multiplication [by $(\sqrt[12]{2})^N$].[5]

In Appendix 12.2 we provide a program to experiment with this sort of synthesis. A note (from an instrument) can be digitized and analyzed to obtain its harmonic structure. The harmonic elements (including 30-40 sidebands) may be captured and saved, and then used to generate simple songs. The program provided isn't very elaborate—it generates sounds over only one octave and doesn't really synthesize musical sounds from an algorithm; however, it does allow the analysis of musical sounds, and it does allow generation of simple songs to illustrate the basic technique. It also allows summing files together so that more complicated pieces could be assembled, but creating synthesizers is not our objective here. We are only illustrating the potential of the FFT to provide high quality sound in the synthesis of virtually any instrument (or even an orchestra). This is the technology we referred to at the end of the last

---

[5] Strictly speaking even this is true only for tones that are near each other; for, as we move up and down the scale on any instrument, the tone changes. As noted earlier this implies a new harmonic model (for any instrument) must be created for every octave (at least). Alternatively, we could *map* the harmonic structure of an instrument over its tonal range and create a dynamic model which changes with *every* note.

Playback Rate

chapter. It's the next technological step in *synthetic sound* (i.e., to improve on the fidelity of recorded/reconstructed sounds—and to recover high fidelity replicas of the very early attempts to record sound).

There is much to do here. We need to understand the structure of the harmonic element side bands. Surely the modulation envelope determines a significant part of the sidebands, but the asymmetry of these side bands tells us there's considerably more here than just the modulation of a harmonic carrier sinusoid. We need to know how distinctly different sounds are carried in these side bands (e.g., in the case of vocal sounds how the vowel sounds are carried via these sidebands). We also need to know how the individual harmonic elements combine to produce the distinctive sounds of various instruments, voices, etc.

In this project it's clear the Fourier transform will be both an instrument of analysis *and* synthesis. We have included a routine to perform a *fine line* analysis of the sounds which employs a DFT fractional frequency algorithm. It's slow in execution, of course, but once we have determined the location of the major harmonic elements we need not analyze a broad band of frequencies. This is one of the major themes of this book—if you want to use the tool of Fourier analysis in practical research and development you will need to improvise, design, and write variations of the FFT (and DFT) which can only be done if you *understand* these tools.

If we can build a synthesizer that *generates* a truly high fidelity replica of musical instruments, we can eliminate the insurmountable problems of trying to *record* live performances. Such a synthesizer would create the music note by note, which allows control of the piece at the time of creation (just as with a live performance). This feature, as we said before, would eventually allow the *real time* synthesis of a high fidelity orchestra, but understanding the nature of sound itself will open whole new fields in the 21st Century.

CHAPTER XIII

THE TWO-DIMENSIONAL FFT

13.0 INTRODUCTION

The two-dimensional transform opens up a whole new world of exploration. It will be easy enough to explain the algorithm and software, but it will take a little longer to explain the "meaning and purpose of existence" stuff; so, we will postpone all that for a while. We will begin with the technicality of writing two-dimensional FFT programs.

13.1 WHAT ARE WE DOING HERE?

When we perform the *one-dimensional* DFT/FFT we are finding the sinusoids which, when summed together, recreate the data points of the original signal. Fig. 13.1 shows a typical example of the conventional one dimensional transform.

Fig. 13.1 - 1-D Transform

In a *two-dimensional* data array (e.g., a digitized picture) we have a *matrix* of data points. If, for example, we take a picture of a star through a good, properly aligned (but slightly out of focus) telescope, we will get something similar to Fig. 13.2A. If we use a CCD camera to take the picture we will actually have an array of numbers which might be displayed as in Fig. 13.2B. Now, we may view this array as horizontal lines of data and, using our familiar 1-D transform, take the transform of each of these lines separately (as shown in Fig. 13.1). There is nothing earth-shattering here—we would simply have a lot of transforms of

Image

Fig. 13.2A

Digitized Data

Fig. 13.2B

Two Dimensional FFT

individual data lines. If, however, we replaced each data line of the original picture with its transform (as we normally do), we would then have a new array of data such as is shown in Figure 13.3.

Let's pause here and talk about what we've done so far. We now have transforms of every line of the original picture (which is, incidentally, the picture of Fig. 13.2). It's apparent we had no good reason for taking the transforms of the horizontal lines—we might just as well have taken transforms in the vertical direction—the relationship of digitized data points (frequently called *pixels*) is just as important in the vertical as the horizontal.[1]

Fig. 13.3 - Row Transforms

Unfortunately, our transformation of the horizontal rows ignores the significance of the vertical relationship (and, of course, if we do it in the vertical direction we lose the relationship in the horizontal direction). This is, of course, a serious problem—to get a complete transform of the picture we must include both vertical and horizontal relationships.

Okay, then, this is the question we are trying to focus on: how can we get a proper transform of the 2-D picture? After taking horizontal transforms, how can we go back and include the vertical transforms? How do we combine vertical and horizontal into a single transform in such a way that the final result makes sense?

Well, in Fig. 13.3 we still have a matrix of data points—we have the phasors (which represent the amplitude/phase of the sinusoids that make up the horizontal rows). Let's look closely at these rows of data. Along the left-most *column* we have all of the *zero frequency* components of the horizontal transforms. The next column over contains all the *fundamental* frequencies of the horizontal transforms, the next contains the 2nd harmonics, etc. Now, there's no reason why we can't consider these columns of *same frequency* components to be *functions* in their own right, but before we willy-nilly start taking transforms in the vertical direction we need to look at what, exactly, these *vertical functions* represent. Consider the left-most column of Fig. 13.3 for example (i.e., the zero frequency components of the horizontal lines). These particular components represent the *average value* of the *intensity* of each line.

---
[1] This is, in fact, a single function—a two-dimensional function—so we need a true two-dimensional transform.

13.2 THE TWO-DIMENSIONAL TRANSFORM

The zero frequency component is obtained by adding all the data points of the horizontal line and dividing by Q (Q = number of data points in a line). Now, this average value is a quantity that applies to the whole line. If we reconstruct the data points of any horizontal line by summing in the harmonics of its Fourier series, this average value is added to each and every data point of the reconstruction. *This average value is a quantity representing the whole horizontal line*; but then, when we think about the other harmonics, their phasors represent *sinusoids* whose amplitudes are also "constant" across the whole line. Each and every one of the sine and cosine *coefficients* tell us the *multiplied amplitude* of that sinusoid across the whole line. So, then, if we go down any of these columns (of specific frequency), we obtain a function representing *average values* of the horizontal lines. The left-most *column*, for example, is a *function* which tells us how the *average horizontal* picture line *magnitude* varies in the vertical direction.

Now, each element of any horizontal transform is a single phasor, and when we consider a vertical column at any frequency, there's no reason we can't take the Fourier transform of that column. The transform of any vertical column will apparently be another frequency domain function—a spectrum of constant amplitude harmonics which may be combined to reconstruct the vertical column again. That is, the components of each vertical transform are sort of *average values* of the vertical lines of data, which, in turn, are sort of average values of the horizontal lines of data. We therefore successfully combine the horizontal and vertical information of the original picture into each point of the 2-D transform. *There's more (obviously), but that's enough for now...we'll come back to this shortly.* Our immediate objective is to write a 2-D FFT program, and we have already uncovered how to do that. This will not be difficult. We will use our *standard* algorithm to take the row and column transforms of a two-dimensional image (i.e., we will revert back to the version that includes negative frequencies) for reasons that will become obvious.

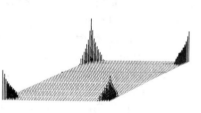

Fig. 13.4 - 2-D Transform

Two Dimensional FFT

13.3 THE 2-D ALGORITHM

First, we must now dimension the arrays for two-dimensional data (but note that the twiddle factors are still only one-dimensional—we only take the transform in one direction at a time):

```
20 DIM C(Q, Q), S(Q, Q), KC(Q), KS(Q), DA1(Q, Q)
```

The old standby FFT routine is located at line 200, but to perform the transform in 2-D we must take the FFT of both rows and columns. That is, after performing all the horizontal transforms, we then transform each column [note: when we do the *inverse* transform we must transform the columns first—then the rows—but, we'll talk about that later). We need two separate routines to handle these forward/reverse and row/column requirements. The high-level forward routine (line 100) looks like this:

```
REM *********************************************
REM *                XFORM FUNCTION              *
REM *********************************************
100 CLS : K6= -1: SK1= 2: XDIR= 1: T9= TIMER 'XDIR: 1=FWD, 0=INVERSE
102 GOSUB 200 ' DO FORWARD ROW XFORMS
104 GOSUB 300 ' DO FORWARD COLUMN XFORMS
106 T9 = TIMER - T9 ' CHECK TIME
108 RETURN
```

We clear the screen, then set the sign constant $K6 = -1$ (for a *forward* transform)—the scale factor constant $SK1 = 2$ (again, forward transform)—the *xform direction flag* $XDIR = 1$ and T9 gets the starting time. At line 102 we jump to the *row* FFT routine:

```
REM *********************************************
REM *                 TRANSFORMS                 *
REM *********************************************
200 CLS : KRTST = 19
202 FOR KR = 0 TO Q1 ' XFORM 2D ARRAY BY ROWS
204 'IF XDIR = 1 THEN GOSUB 400
206 PRINT USING "###_ "; KR; ' PRINT ROW # BEING XFORMED
208 IF KR = KRTST THEN PRINT : KRTST = KRTST + 20' END PRINT LINE
    '    *********************************
    '    * THE ROUTINE BELOW IS FOR A ROW *
    '    *********************************
210 FOR M = 0 TO N1: QT = 2 ^ (N - M)' DO N STAGES
212 QT2 = QT / 2: QT3 = QT2 - 1: KT = 0
214 FOR J = 0 TO Q1 STEP QT: KT2 = KT + 1' DO ALL FREQUENCY SETS
216 FOR I = 0 TO QT3: J1 = I + J: K = J1 + QT2' DO ALL FREQUENCIES IN SET
    ' ROW BUTTERFLY
218 CTEMP = (C(KR,J1) + C(KR,K)*KC(KT) - K6*S(KR,K)*KS(KT))/SK1
```

```
220 STEMP = (S(KR,J1) + K6*C(KR,K)*KS(KT) + S(KR, K)*KC(KT))/SK1
222 CTEMP2 = (C(KR,J1) + C(KR, K)*KC(KT2) - K6*S(KR,K)*KS(KT2))/SK1
224 S(KR,K) = (S(KR,J1) + K6*C(KR,K)*KS(KT2) + S(KR,K)*KC(KT2))/SK1
226 C(KR,K) = CTEMP2: C(KR, J1) = CTEMP: S(KR, J1) = STEMP
228 NEXT I' ROTATE AND SUM NEXT PAIR OF COMPONENTS
230 KT = KT + 2
232 NEXT J' DO NEXT SET OF FREQUENCIES
234 NEXT M' DO NEXT STAGE
    ' BIT REVERSAL FOR ROW TRANSFORMS
236 FOR I = 1 TO Q1: INDX = 0
238 FOR J = 0 TO N1
240 IF I AND 2 ^ J THEN INDX = INDX + 2 ^ (N1 - J)
242 NEXT J
244 IF INDX>I THEN SWAP C(KR,I), C(KR,INDX): SWAP S(KR,I), S(KR,INDX)
246 NEXT I
248 'IF XDIR = 0 THEN GOSUB 400
250 NEXT KR
252 T9 = TIMER - T9: GOSUB 176' SHOW RESULTS OF ROW XFORMS
254 A$ = INKEY$: IF A$ = "" THEN 254
256 CLS : T9 = TIMER - T9: RETURN' ROW TRANSFORMS DONE
```

At line 200 we define a constant KRTST = 19. This has nothing to do with the transform but on larger arrays it may take a while to perform a transform, and we need something to indicate the computer is still running; therefore, the number of each row is printed as it is transformed (line 206). KRTST is used to control the printed line length (line 208). We will explain why line 204 is presented as a comment shortly.

At line 214 we come to the familiar FFT routine. We process the row of data as we would in any FFT, emerging from this *row transform routine* at line 248, and looping back for the next row at line 250. When all rows are transformed we display the data (line 252) and then return to the high level routine (line 104) [where we will immediately jump down to transform the data by columns]. The difference between row and column is the reversal of row and column addressing in the butterflys (lines 218 thru 226). The column routine is shown in the appendix (starts line 300).

There are differences, of course, in the routines to generate 2-D data, plot the data, etc., but we discuss all of that in Appendix 13.1 (where a complete listing is provided).

13.4 FUNCTIONAL AND OPERATIONAL CONSIDERATIONS

First of all, you will only be able to implement an array of 64 by 64 before you run out of memory. If you compile this program (setting the /AH "switch" when starting QuickBasic™) you can extend this to 128 by

Two Dimensional FFT 163

128, but that's as far as you will be able to go in this Basic language (without writing software to shuffle data back and forth to the hard drive). The reason is that DOS has always been hamstrung by the original (in my opinion poorly conceived) addressing scheme of the 8080 µprocessor, which basically provided for one mega-byte of memory. Today DOS requires more than half of its one meg. When you load the application you want to run (QuickBasic™ for example), that takes up a big chunk of what's left, and the program you write in QuickBasic™ takes up another chunk, and you wind up with only a few hundred kilo-bytes left over to hold data. On my computer, when the program FFT13.1 is loaded and running [you can add the line of code PRINT FRE(-1) both before and after dimensioning the arrays], there is 329,960 bytes available for data arrays (this number is affected by such remote things as the number of BUFFERS in the CONFIG.SYS) file. Now, floating point (i.e., real) numbers take 4 bytes of memory each, and we have two 2-D arrays of data (one for the cosine component and another for the sine). There are also two smaller arrays (Q quad-bytes for each twiddle factor array) and a "spare data array" of QxQ quad-bytes, which adds up to:

Total memory = $(2 \cdot Q^2 + 2 \cdot Q + Q^2) \cdot 4$ bytes.
$= Q^2 (3 + 2/Q) \cdot 4$ bytes $\approx 12\ Q^2$ bytes

If Q = 64 this will require 49,664 bytes; but, when Q = 128 this demands 197,632 bytes. Q = 256 requires 788,480 bytes so, if you want to work with arrays larger than 128 by 128 pixels, you will have to resort to more desperate techniques—sorry about that.

Execution times have become a moot point. Modern computers are so fast that lousy system software doesn't matter for the above array sizes. On a 500 MHz machine, you can do a 128 x 128 transform in about 10 sec; still, if we could see things clearly, we would see that, rightly, the value of software (assuming it works) is *inversely* proportional to: 1) the execution time, 2) number of keystrokes (or mouse operations) required to perform its tasks and, 3) quantity of code. When we pay hundreds of dollars for a piece of software, however, we want it to contain a *lot* of code that takes a lot of time to load and run...and the software companies oblige us with stuff that's optimized primarily to make a profit.

The last thing we need to talk about is the frequency domain arrangement that results from the conventional 2-D transform. In Figure

13.4 we see the transform comes out in four *heaps* in the corners of the plot...and some see this to be a gross distortion of reality. As everyone knows, the negative part of the X axis should be to the *left* of zero while the positive should be to the right (just as *North* should be *up* on a map), and the positive Y axis should be *above* the X axis (*East* should be to the right). As we all know, when we take the conventional transform, we get the negative frequencies above the positive ones (i.e., they are plotted to the right of the positive spectrum) and, if you review the mechanism described in Section 13.2 of this chapter, you will see how the negative frequencies come out "right and below" the positive components—but running in the wrong directions from each end. Now, if you can mentally shift the quadrants around where they *should be* (the preceding comments will be an immense help here), you will see the four heaps are really part of the same thing—a single heap in the center (Fig. 13.5)!

Okay, we could go all the way back to the original DFT program, write it such that the negative frequencies come out at the bottom, and then develop our FFT algorithms to give the same results; but, that would be an awful lot of (painful) work. There's another, relatively simple correction that we can introduce here—one that will straighten things out....

In Chapter 10 we noted that if the sampling pulse train was composed of alternately positive and negative pulses, the required sampling rate would then be *equal to* the Nyquist frequency in the signal (Section 10.8). We also saw this eliminates the zero-order harmonics, puts the negative frequency components in that space, and the positive frequencies will therefore be above the negative frequencies. Now, if we go through the data and negate every-other data point, we will have accomplished this sort of 1:1 sampling—and transforming this data will then yield a spectrum with the positive and negative frequencies in correct relative position. If we take the sampling frequency as the zero frequency in this 1:1 modulation scheme, everything will work-out in our graphics.

```
' *******************************************
' *           MODIFY ROW SAMPLING           *
' *******************************************
400 FOR I = 1 TO Q - 1 STEP 2
402 C(KR, I) = -C(KR, I): S(KR, I) = -S(KR, I)
404 NEXT I
406 RETURN
' *******************************************
' *          MODIFY COLUMN SAMPLING         *
' *******************************************
```

Two Dimensional FFT

```
410 FOR I = 0 TO Q - 1 STEP 2
412 C(I, KR) = -C(I, KR): S(I, KR) = -S(I, KR)
414 NEXT I
416 RETURN
```

We will need to add code in the transform routine to jump to these routines before performing the transform—we need to modify the sampling prior to performing the FFT:

```
204 IF XDIR = 1 THEN GOSUB 400
```

We also need to add this prior to each column FFT:

```
304 IF XDIR = 1 THEN GOSUB 410
```

If we are doing a *forward transform* we jump down to the modify sampling routines but we must perform the *reverse* modification on reconstruction. Furthermore, when we reconstruct the signal, we must perform the inverse transform *before* we modify the sampling—that will properly *undo* what we *did* during the forward transform. The lines of code required are:

```
348 IF XDIR = 0 THEN GOSUB 410
```

On reconstruction we undo the *columns* first. Also note that, in either case, we use the same routine—negating every other data point twice yields the original data. We then add the final *undoing* of the data:

```
248 IF XDIR = 0 THEN GOSUB 400
```

which should get us back to the original format of the input data. You can add these lines of code to the program of Appendix 13.1 by simply removing the apostrophe before the lines noted, and the transform will appear as in Fig. 13.5—as opposed to Fig. 13.4.

Fig. 13.5 - Modified Sampling

CHAPTER XIV

TRANSFORM FAMILIARIZATION

14.0 INTRODUCTION

In this chapter we delve into the *nature* of the 2-D transform. We pick this up where we left off in the middle of the previous chapter, and will find that, for the most part, the nature of the 2-D transform is just as intuitively obvious as the 1-D transform. Nonetheless (as you probably suspect), there's enough novelty to make this interesting—even fun.

14.1 THE 2-D HARMONICS

Now, as we said (Section 13.2), when we take the transform of the horizontal rows, the *zero frequency components* are obtained by summing all the pixels in the row and dividing by Q, which represents the average value of the function. So, when we take the transform (going vertically) of all the zero frequency components (i.e., the left-most column of the horizontal transforms), we obtain another spectrum of frequency components. The zero frequency component of *this* vertical transform will be the summation of the zero frequency data points of all the horizontal transforms (i.e., a summation of the average values of the horizontal rows) divided by Q (again). This zero frequency component is the average value of *all the pixels*—the average intensity of the whole picture—it is the "zero, zero" component of the 2-D transform.

Okay, the zero, zero frequency represents the average intensity of the picture, but what about the other harmonics of these vertical transforms? Well, take a look at Fig. 14.1—this is the digitized data of a picture of horizontal sinusoidal "bars" (with the sinusoids running vertically). Now, each horizontal FFT will only see a constant value, so there will be no real sinusoid components—only the zero frequency—the "average value." The horizontal transforms will yield the data shown in Figure

Fig. 14.1 - Vertical Sinusoids

Transform Familiarization

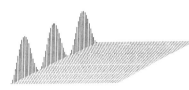

Fig. 14.2 - Row Transforms

14.2. This is very interesting! It shows that the average values of the horizontal rows vary in a *sinusoidal pattern*! Who would have guessed? And what will the vertical transforms yield?

In Fig. 14.3 we see the final "2-D" results of the transform for Fig. 14.1. There were three cycles of the sinusoidal horizontal bars and, sure enough, there is a single component at the 3rd harmonic (and, of course, its negative frequency counterpart at the other end). Now, we have discussed how the vertical FFTs take the transform of the horizontal *average values*, and here we see what that means. If you have a vertical sinusoidal pattern in your picture it will show up as a pronounced frequency component in the *vertical zero frequency column* of the final transform. But we usually expect more than just vertically running sinusoids in our art work, so let's continue our investigation.

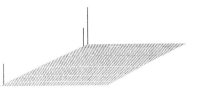

Fig. 14.3 - 2-D Transform

In Figure 14.4 we have turned the pattern of Fig. 14.1 by 90°. By now the results of the *row transforms* of this data matrix should come as no surprise. In Fig. 14.5 we see each horizontal transform yields a zero component and a third harmonic component. Big deal, we could have predicted *that* by inspection. Similarly, there's no variation along any of these columns, we will only get zero frequency components in the vertical transforms. Indeed, Figure 14.6 shows that is precisely the case. Horizontal sinusoids have 2-D transforms composed of a zero, zero frequency component (i.e., the average value of the picture), and a zero, nth harmonic component at the frequency of the sinusoid.

Fig. 14.4 - Horizontal Sinusoids

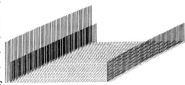

Fig. 14.5 - Row Transforms

Apparently, the *zero frequency* rows and columns of a 2-D transform indicate perfectly horizontal and vertical sinusoidal patterns in the picture.... But what about all the other components—the ones that are not in a zero frequency row or column (i.e., the *vast* majority of

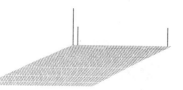

Fig. 14.6 - 2-D Transform

the harmonics)? Well, we just saw that the zero frequency *column* implies there is *no* variation of the signal across the *row*, while the zero frequency *row* implies there's no variation of the signal along the *column*; so, right away we would suspect the components to the right of the zero column, and below the zero row, would vary in both directions. Components in the 1st column (to the right of zero) imply a single cycle of signal *across the row* (indeed, that's what Fig. 14.6 shows us). Components in the 1st *row* imply one cycle of signal *down the column* (e.g., Fig. 14.3).

So, suppose we generate the same three cycle pattern used in Fig. 14.1, simultaneously making it vary by three cycles across the row. In other words, we sort of combine the patterns of Fig. 14.1 and 14.4 as shown in Fig. 14.7. *[Note: this two-dimensional sinusoid is the sum of two variables*:

Fig. 14.7 - 3 x 3 Sinusoids

$$f(S) = Sin^2(\pi(3x + 3y))$$

which results in a diagonal pattern of Fig. 14.7. Note that there are 3 sine waves clearly visible on both the x and y end patterns]. The horizontal transforms (Fig. 14.8) come out pretty much as we expect. *[Note: we have*

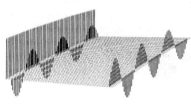

Fig. 14.8 - Row Transforms

fudged this result for the row transforms by showing only the sine component (to illustrate our point). The actual *magnitude of any row transform is the RSS of both sine and* cosine *components which yields a constant magnitude for the column of 3rd harmonic components.]*

Transform Familiarization 169

So, all of the components in a 2-D transform (other than the components in the zero column or row) imply sinusoids oriented at an angle to the x and y axes. The frequencies of the components along the x and y axes determine the number of cycles along those axes.

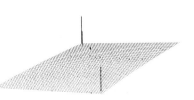

Fig. 14.9 - 2-D Transform

It should be apparent that this is the same interpretation of the 2-D transform given in Chapter 13. The implication here, as in the 1-D transform, is that we can reproduce any real picture via a summation of these 2-D sinusoids. *[The textbooks prove this, of course, but let's not be too dense here. If the horizontal transforms can reproduce the row data and the vertical transforms can reproduce the horizontal transforms, we can surely reproduce the picture; however, we have just seen that any one of the 2-D components reconstructs to a 2-D sinusoid....]* We will give another illustration of this mechanism a little later; but, at this point, you should be confidently aware that you know precisely how to generate any one of the 2-D harmonics and, conversely, exactly what any one of the harmonics will produce in the reconstructed picture.

14.2 THE *VERY* DREADED SINC^2 FUNCTION

[We used the conventional display for frequency domain in the previous section (i.e., negative frequencies to the right of the positive ones) so we could draw on our familiarity with the 1-D transform. We now change back to our modified 2-D algorithm and display the frequency domain as shown in Figures 14.10 and 14.11 (i.e., with the 0,0 frequency at the middle of the chart). We no longer use the 1:1 sampling technique, however—we use quadrant switching as explained in Appendix 14.1]

So, the 2-D transform is not really different from the transform we have been studying—we have only extended its application to a second dimension. We would expect most of the considerations we have discussed so far to carry through to this extended application; however, we also anticipate significant differences in the 2-D phenomena—and we need to be aware of these differences. One prominent example is the ubiquitous *sinc function* we have discussed so extensively. We have included the 2-D version in the *Function Generator* sub-routine, and if you select that function, you will obtain the image we exhibited in Fig. 13.2B.

The transform of the sinc² function is shown in Fig. 14.10 (which is the same transform shown in Figures 13.4 and 13.5). So, what's the significance of this vaguely familiar function? You recall, in Chapter 10, we found the sinc function was generated (as sidebands) on every transform that begins and ends abruptly (i.e., almost every DFT/ FFT). That is, the sinc function is the transform of a rectangle function, and since a rectangle necessarily *multiplies* all time domain data we analyze, its transform is *convolved* with the frequency domain data. *[As you also recall, we don't see the sinc function in the conventional transforms because it happens to go to zero at every data point; nonetheless, we know if we look between the data points of a conventional transform we will find it there. In short, if we abruptly truncate the time domain data (as in a fractional frequency transform) we see the sinc function in the frequency domain, and when we truncate the data in the frequency domain (à la Gibbs phenomenon) we get a sinc function in the time domain.]*

Very well then, in our present case we are dealing with two-dimensional functions, and we might imagine that a sinc function would still be generated in the transform domain for all functions that start and stop abruptly. The problem here is that, in the 1-D FFT the ever-present sinc function results because we *implicitly* multiply the time domain function by *unity* over that part of the signal we analyze (and zero everywhere else). That is, we don't *change* the amplitude of the signal inside the *aperture*—we multiply every data point by unity. [In the illustration above, however, we generated the 2-D sinc² in the space domain because, as we shall see, it's the frequency domain that's *bounded*.]

Fig. 14.10 - 2-D Xform

Now, as any fool can plainly see, the problem here is that the *sinc²* transform (Fig. 14.10) is *not* flat in the frequency domain! It *peaks* at the zero, zero frequency and then rolls off until it comes to the limiting aperture where it dives into oblivion; but multiplication by a rectangle is only *implicit!* It only comes about because real data always has a beginning and end. The sinc² function transforms to something that looks like a "kiss" from Hersey, PA...and that's obviously not what we're talking about.

14.3 THE VERY *VERY* DREADED BESSEL[1] FUNCTION

We want a function that doesn't distort the signal within the aperture, we need a function that multiplies all frequencies equally until we come to the boundary of the aperture—some function that multiplies all harmonics *within* the aperture by unity (and by zero outside)—it being more or less obvious there *must* be a function that does this.

If you are familiar with optics you know that the function we seek does indeed *look like* the $sinc^2$ shown in Figure 13.2; but, there are a lot of functions that have this general appearance. One prominent candidate is the Bessel function (of the first kind of zero order). From a casual glance it looks a great deal like the $sinc^2$ function and its transform is much closer to what we desire (Fig. 14.12); but, it's still not perfectly flat across the aperture—it's still not *precisely* what we want.

Fig. 14.11 - Bessel Function

Fig. 14.12 - Bessel Xform

BESSEL II

Nothing less than a perfectly flat-topped function will do; but how shall we find it? Well, we know how to generate individual harmonic components—we learned how just two sections back using *sinusoidal bars*.

We can use a technique similar to the one we used back when we first discussed Fourier series, and write a routine to sum sinusoidal *bars* at the frequency components where we want them—and leave zeros where we don't. That's how we do it in the Bessel II routine in FFT 14.01.

Fig. 14.13 - Bessel II Xform

---

[1]Friedrich Wilhelm Bessel (1784-1846), a contemporary and friend of Gauss. Bessel was the first to measure the distance to a star (61 Cygni) using parallax due to the Earth's orbit about the Sun. He made many other significant contributions to astronomy; however, he is probably best known for these *radial* functions.

You will note that Fig. 14.13 is indeed what we are looking for (it could hardly be otherwise), but is Fig. 14.14 what we see in our telescopes?[2] [*Note:* This is not the best way to generate this particular function—it just illustrates that we can create a modestly complex picture by piecing together sinusoidal bars (see discussion p. 169 and Bessel II in Appendix 14.1). This is obviously a legitimate interpretation of the 2-D transform.]

Fig. 14.14 - Bessel II

There's a much easier way to find this *aperture function*. We can simply generate a constant amplitude function over the desired aperture *in the frequency domain*, take the inverse transform, and— voilà! We have also included *this* in our function generator sub-routine (Bessel III). It's easy to generate, illustrates the same thing as all those 2-D sinusoids, and represents what we really want. [At this point you should probably load program FFT14.01 into your computer and experiment with these functions a little.]

14.4 ANOTHER IMPORTANT 2-D FUNCTION

There's one last function we need to consider in this development of the *aperture function*—the *star* (i.e., impulse function). Stars are so distant they make excellent optical point sources. From such great distance all of the rays of light arrive virtually *parallel*, and they serve very well to test such things as a telescope's primary lens (i.e., objective).

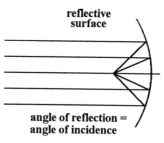

Fig. 14.15

Geometrically, if the surface of a Gregorian objective has a perfect parabolic form, and all the rays falling on the objective are parallel, they will converge to a single point. This would be an equal intensity illumination (via parallel rays) across the whole surface of the objective—and the lens *transforms* this *equal intensity illumination at its*

---

[2] One way to see this is to cut a small hole in the lens cap of your refracting telescope (using, for example, a leather punch) and project a laser beam (expanded to a 1" to 2" diameter) onto this "stopped-down" aperture. The bright spot at the center of this pattern is know as the *Airy disk* (for Sir George Biddell Airy, 1801-1892). It's amazing how few people have actually seen this pattern, but *you* can purchase a "key-chain" laser for about $10.00 (target the discount stores—automotive accessories—or Edmund Scientific).

Transform Familiarization 173

aperture to a *point* at the focal plane...and from all our foregoing discussions we recognize this point to be a 2-D impulse function. Now, from our earlier discussions of impulse functions, we know their transforms are constant amplitude cosine harmonics, and you might anticipate that the 2-D transform of this *star function* (i.e., impulse function) will generate perfectly equal amplitude harmonics across the whole 2-D *frequency* domain. That is, the Fourier transform of this *star* is very much like the relationship between the image created by a lens and the equal intensity parallel rays that fall on the lens aperture...we'll certainly have to discuss *this* further!

14.5 2-D SIMILARITY

It's an interesting exercise to generate a series of increasing diameter apertures and observe the 2-D *similarity* theorem on reconstruction (using the Bessel III functions). The *image* expands and contracts inversely proportion to the expansion and contraction of the frequency domain aperture (as we all surely expected); however, in this 2-D version, we need to point out explicitly what this means for an image. If a *star* function is nothing more than an impulse function—a single data point—and the transform of this point is a completely flat spectrum, then if we limit the aperture (in the frequency domain by zeroing-out all of the components beyond some arbitrary *radius*) we will have the very same frequency domain function that we used (see Fig 14.13 on *p*. 171) to generate our Bessel III function.

So, then, if we start with a point source, transform to the frequency domain, limit the aperture, and transform back to the space domain, we will have turned our point into a Bessel function! Furthermore, the smaller we make the aperture, *the more* spread-out *the Bessel function becomes!*

Here we must point out that, when we limit the aperture in the frequency domain, we are doing the same thing we did when we illustrated the Gibbs Phenomenon—or the problems associated with oversampling—we are looking between the pickets of the picket fence again! In this case, where we are dealing with a 2-D transform, we find the summation of sinusoids (i.e., the mechanism of the FFT) yields a variation on a Bessel function (as opposed to the familiar *sinc function* of the 1-D transform).

In real optical systems, the image we get from finite aperture objectives is indeed limited by this Bessel function—we are doing the same thing in both cases (see footnote opposite page). We will discuss this in greater detail in the next chapter.

CHAPTER 15

INTRODUCTION TO FOURIER OPTICS

Geometric Optics is based on the fact that light travels in straight lines (see Fig 14.15) with the notable exceptions that it is bent by transparent media (e.g., it's refracted by water, glass, the atmosphere, etc.) and also reflected by smooth surfaces. Now, as we examine phenomena such as refraction and reflection (and especially image formation) it becomes apparent we don't really understand optics until we understand the *wave nature* of light—and as we study the mechanics of these phenomena via waves, we discover an astonishing relationship to the Fourier transform.

15.1 WAVES

When a wave rolls in to the shore it propagates only in the forward direction. We might guess that the forward motion of the wave is due to the weight of the water piled up at the crest trying to fall back down to the level of the surrounding sea, but this falling-down process never quite ends—apparently something in the propagation mechanism continues to pull water upward into the oncoming wave and, even though the wave propagates forward, the piled up water *falls* down 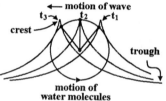 the back side of the wave. We will not go into this particular phenomenon here but, rather, consider a peripheral characteristic.

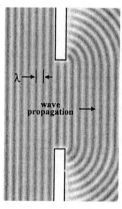

Now, surely, the water that's raised up to the crest of the wave experiences a gravitational force downward, and the only thing that keeps this water from falling down in the *side-wise* direction is the countering force generated by an equal height of water to each side of any point along the wave's crest. In fact, if a wave passes through an opening in a sea-wall, for example, we find that water does indeed start falling down in a sidewise direction at each end of the wave (after passing through the opening), and this is the mechanism we will consider.

Wave Optics Mechanism

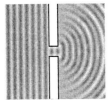

As the opening in our sea-wall is made smaller we come to a limiting case—the wave propagates beyond the hole with a circular wave front. The great Christian Huygens (hoi´-gens or hi´genz(?)—1629-1695) recognized that, if the principle of superposition holds, then every point on a wave front may be considered as a source of these circular propagating waves. For waves of reasonable breadth, the symmetry of these side-by-side point sources will cancel out the side-wise propagation, leaving only a forward propagating wave; however, at abruptly ending boundaries the asymmetry exposes the sidewise, circular, propagation. This viewpoint helps explain the phenomenon of *diffraction*—a form of *interference*.

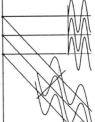

15.2 INTERFERENCE

If two waves propagate through the same region, there will be places where the crest of one wave coincides with the *trough* of the other. At these places the crest will cancel the trough (i.e., the resultant water level will neither be high nor low) and the effect of gravity will be nullified. We might imagine that once the two waves cancel, propagation would terminate; however, that's not the case. The two waves do indeed cancel where trough and crest coincide, but farther along we find both waves propagating just as they did before.[1] Clearly this phenomenon wouldn't take place if superposition didn't hold for propagating waves. Furthermore, any time we observe an interference pattern we might well imagine it's a phenomenon arising from an underlying wave motion (i.e., the interference pattern of two superimposed *light waves* strongly suggests that light propagates as waves.[2])

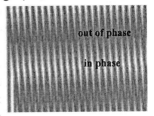

---

[1] Waves are not simply static variations in the height of the water, but dynamic interchanges of potential and kinetic energy. The phenomena of superposition/interference clearly expose this *dynamic* characteristic.

[2] The cycling of *potential energy* (e.g., the *height* of the water in a gravitational field), and the cyclical reversing of *kinetic energy* (e.g., the back and forth motion of water in the wave) are, as abstract phenomena, the fundamental characteristic of all wave forms.

15.3 REFRACTION AND WAVEFRONTS

Refraction is the phenomenon of *light-bending* at boundaries of transparent substances. This is due to the change of velocity as light moves from one material to another (and the wave nature of light). To understand this phenomenon we must clearly understand the concept of a *wave-front*.

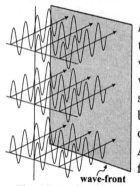

If you consider the cross-section of a *key-chain laser* beam as it propagates across the room, you will have the basic idea of a wave-front. That is, light propagates like the waves you make on a garden hose when you shake one end up and down, but since the beam has a finite cross-section we must think of many ropes (in parallel) all carrying waves. All of these sinusoid *elements* combine to form a single wavefront.

Fig. 15.6 - Wavefront

Now, in a laser beam, all of the wave elements are in phase, and the beam propagates directly forward. If the beam strikes a plate of glass at an angle, however, the bottom of the beam strikes first, and progressively higher parts of the wavefront strike the glass at progressively later times (and points). Since light moves slower through glass (than air) the bottom of the beam (in glass) will slip behind the parts of the beam that are still propagating in the air, and the phase of the wave front is successively advanced as it passes into the glass.

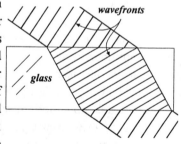

We note that the direction of propagation remains *normal* (i.e., at right angles) to the *wavefront* (as we might anticipate from Huygen's *point source/superposition* principle). This concept of a *wavefront*, while reasonably obvious, is important—it will play a prominent role in our following discussions.

15.4 DIFFRACTION

Now, there's a particularly revealing form of interference that arises when a single beam of light passes through an *aperture* (not unlike the illustration of water waves passing through a gap in a sea wall). If you project your key-chain laser beam through a slit you will find an intensity pattern (projected beyond the slit) of a sinc functions (see Fig 15.17, *p.* 183). This single beam interference is known as *diffraction* and it comes about as follows:

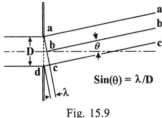

By Huygens' model we know each point on a wavefront radiates forward omnidirectionally. We know in the *near-field* the sidewise components of radiation will cancel; but, recognizing that superposition must hold, it's apparent wave motion at an angle must eventually become separated from the forward moving beam. Now, light radiated *directly* forward will surely move forward as we intuitively imagine it would; but, when we consider radiation at an angle (i.e., other than perpendicular to the plane of the slit) things are not exactly simple. We must look closely at Huygens model to appreciate this effect: at some angle θ to the aperture normal, we have a geometry as shown in Fig. 15.9. If we examine the rays moving at the angle θ we see that the phase along c-c' is delayed by one wavelength (λ) relative to the phase along a-a'. Obviously, the phase along b-b' is delayed by $\lambda/2$.[3] In fact, if we consider small increments across the wavefront moving at the angle θ (Fig. 15.10) we see that the phase along b_4 must be half a wavelength out of phase with a_4, as is the phase between b_3 and a_3, etc., etc. That is, every increment across the wavefront will have an increment that is exactly *out-of-phase* half a slit width (i.e., D/2) away. If all these *out-of-*

Fig. 15.9

$\mathrm{Sin}(\theta) = \lambda/D$

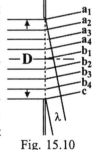

Fig. 15.10

---

[3]With a laser the light is in-phase across the wavefront as it exits the slit. For the ray at c-c', then, light from a Huygen's point source radiator must travel farther to the wave front that is normal to θ (than the ray at a-a').

phase wave front increments are summed together at a point in space, the result will be an interference null (it's an interesting exercise to complete this laser/slit/diffraction illustration)....[4]

15.5 LENSES AND IMAGE FORMATION

In the previous chapter, we noted the function of a lens is to cause all parallel rays (entering the aperture of a lens) to converge to a point. From our discussion of refraction it's more or less obvious that, if we grind and polish a piece of glass to the correct shape, we can make parallel rays striking the glass bend in such a way that they will converge to a point. If we do this for rays that are parallel to the optical axis, then surely the same thing will happen (pretty nearly) for rays that are only slightly misaligned with the optical axis.[5] In

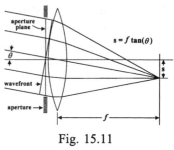

Fig. 15.11

fact, parallel rays entering the aperture will converge (approximately) at a position equal to the tangent of the angle of those rays times the focal length of the lens (see Fig. 15.11). As we also noted, parallel rays are the sort of thing we receive from a distant "point" source (e.g., a star). Now, we can extend this illustration by considering a cluster of stars (e.g., the Pleiades—the star cluster named for the fabled "seven sisters" of mythology depicted in Fig. 15.12).[6] If we focus our telescope on this

Fig. 15.12 - Pleiades

cluster the parallel rays of light from each star will enter the aperture at a different angle and converge at the points shown. We point out again that superposition holds here—the light coming from each star may pass through the lens simultaneously and still be focused into its proper position in the image (we emphasize this

---

[4]That is, as we have left this explanation, the θ directed wave front increments are *not* superimposed. Light, however, is leaving the slit at all angles, of course.

[5]Known as *paraxial* rays.

[6]The Pleiades were the seven daughters of Atlas, the nymphs of Diana (Jupiter turned them into a constellation to aid their escape from Orion); but, we only see six stars. The Pleiad Electra left her sisters, unable to gaze upon the fall of Troy, which had been founded by her son Dardanus (this stuff is important so let's get it right).

Wave Optics Mechanism

superposition holds for light rays at different angles! All rays at the same angle, regardless of where they enter the aperture, will focus into the same image point)! Note that each bundle of rays at any angle will have a wavefront (Fig. 15.11)! The light from each star in the Pleiades will enter the aperture as a wavefront slightly tilted from the others.[7] Let's also point out that, as two points come closer together, their wavefronts approach the same angle, and tend to be focused into the same image point—we will discuss this in more detail shortly.

15.6 THE *2-D FFT* AND IMAGE FORMATION

In FFT15.00 we have a function generator that generates a single star (impulse) which may be placed at any point in the image domain. We will consider a point on the optical axis for our first example (i.e., at position 0,0 in the image domain). Now, from our discussion of impulse functions (back in Ch. 10, *pp*. 120-121) the row transform of a single *star*, being an impulse, will give us a flat spectrum of cosine phasors. Now, when we do the vertical transforms, all of these equal amplitude cosine components will themselves transform as if they were impulses, and we will get a 2-D matrix of equal amplitude cosine harmonics. Furthermore, we know the magnitude of the impulse must be equal to the *square* of the magnitude of each harmonic.

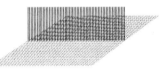

Let's now consider stars that are not on the optical axis. As we know, if we shift the position of our star along the row, the phases of the harmonics will shift proportionally. That is, we will still get the same equal amplitude components in the frequency domain, but now they will

---

[7]This illustration is just a limited example of *all* image formation—every point on the surface of illuminated objects is a point source of light. Note carefully that the information here is not simply the light, but the location and intensity of the light—the information is the *spatial modulation* of the light!

be phase shifted (proportional to the amount of the position shift and harmonic number). It will be easier to demonstrate the impact of this effect in the 2-D transform than to try to explain it.

We have provided a routine (in FFT15.00) to print out just the phases of the 2-D transform. We know the magnitude of harmonics remain constant as the position of the impulse is moved about in the image, so phase is all we need to observe. In the table below we show the phases of the harmonics when the impulse is on the optical axis:

			Col. 0			
0.000	0.000	0.000	0.000	0.000	0.000	0.000
0.000	0.000	0.000	0.000	0.000	0.000	0.000
0.000	0.000	0.000	0.000	0.000	0.000	0.000
0.000	0.000	0.000	0.000	0.000	0.000	0.000
0.000	0.000	0.000	0.000	0.000	0.000	0.000
0.000	0.000	0.000	0.000	0.000	0.000	0.000
0.000	0.000	0.000	0.000	0.000	0.000	0.000

(Row 0 is the fourth row.)

As we expected the phases of all the harmonics are zero—these are all cosine components. If we now shift the star one position along the *row* the phases of the harmonics will be as shown in the table below (their amplitudes remaining constant). We only show the first three rows and columns about the 0, 0 axis but this is sufficient to show what we need to know. Note that all the phase shifts are the same across the *rows*.

-16.875	-11.250	-5.625	0.000	5.625	11.250	16.875
-16.875	-11.250	-5.625	0.000	5.625	11.250	16.875
-16.875	-11.250	-5.625	0.000	5.625	11.250	16.875
-16.875	-11.250	-5.625	0.000	5.625	11.250	16.875
-16.875	-11.250	-5.625	0.000	5.625	11.250	16.875
-16.875	-11.250	-5.625	0.000	5.625	11.250	16.875
-16.875	-11.250	-5.625	0.000	5.625	11.250	16.875

In the next table we show the phase shifts of the harmonic components for a shift of one position along the *column*.

-16.875	-16.875	-16.875	-16.875	-16.875	-16.875	-16.875
-11.250	-11.250	-11.250	-11.250	-11.250	-11.250	-11.250
-5.625	-5.625	-5.625	-5.625	-5.625	-5.625	-5.625
0.000	0.000	0.000	0.000	0.000	0.000	0.000
5.625	5.625	5.625	5.625	5.625	5.625	5.625
11.250	11.250	11.250	11.250	11.250	11.250	11.250
16.875	16.875	16.875	16.875	16.875	16.875	16.875

We get the same results except the wave front now tilts in the vertical direction. Perhaps you're beginning to see where this is heading; so, let's stop skirting the issue and come directly to the point!

Wave Optics Mechanism

15.7 THE FOURIER WAVEFRONT

In the one dimensional transform we do not, in general, suppose the harmonic components exist in some obscure *frequency space*, but recognize that the harmonics are simply superimposed on each other in the *time domain* to generate the time domain signal itself. Indeed, that's the interpretation we have placed on the 2-D harmonic components in the previous chapter—they are simply 2-D sinusoids that are superimposed in the image. Be that as it may, we are now about to propose that, in the 2-D transform, there is another interpretation of the frequency domain— one in which the transform harmonics do indeed exist in another place. This *transform space*, however, is no recondite, ambiguous or obscure place—no 5th dimension or hyperspace. This transform domain is still the 3-D space we normally live in, but a different region of that 3-D space than where the image exists. Let's investigate this new viewpoint.

Back on Page 176 we discussed a particular view of an optical wavefront. We then saw how we could change the direction of propagation by progressively slowing the *sinusoidal elements* of this wavefront (i.e., refraction), and how diffraction came about when these *elements* interfered with each other. Now, we have just seen how any pixel in an image (i.e., an impulse function) transforms to a function of equal amplitude sinusoids, and how the phases of these sinusoids depend on (and determine) the position of the pixel in the image (see previous page tables). Note that the picture presented by these tables are exactly what we described as the tilted wavefronts entering a lens! The uniformly changing phase shift across the transform plane is precisely what we described as a tilted wavefront entering a lens, or the self interfering wavefront generated when light passes through a slit. The transform of any pixel is a matrix of phasors (i.e., sinusoidal *elements*) which describe the wavefronts shown in Figs. 15.6, 15.10 and 15.11. But, even though the transform of any pixel corresponds to an optical wavefront, the routes by which we arrive at optical wavefronts and transform planes are quite different! Are these really the same?

First of all, in the FFT we do not deal with light, but with the spatial modulation of light...let's look at this carefully. A lens bends the parallel rays of light entering the aperture in such a way that they converge to a single point in the image; but, in the FFT we derive the transform (i.e., the spatial frequency version of the aperture plane) by extracting *different frequency* components from the image. It's not clear at this point why these two processes arrive at the same end result, or even that the results are really the same. We must keep in mind that the FFT is *not* modeling the physical mechanics of an optical system—it's describing another process, another mechanism; nonetheless, these mechanisms are so intimately connected they *corresponds* to the same thing. That is, under the supposition of a sinusoidal basis function (i.e., waves) for both light and the Fourier transform, things like diffraction and refraction through a lens are equally captured by both processes. A specific example might be of help here:

The components near the outer periphery of a transform are the high-frequency components which provide sharp definition in the image, but it's not clear that rays through the outer periphery of an aperture contribute more to sharpness and definition than rays on the axis. Are these *really* the same?

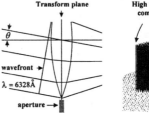

As the Airy disk grows smaller, definition and sharpness increase. A careful review of Figs. 15.9 and 15.10 will reveal that as the aperture increases, the angle of the null point decreases, and the diameter of the Airy disk decreases. Now, we can surely write an equation for the diameter of the Airy disk vs. the diameter of the aperture (see Fig. 15.9)—and just as surely we can find an equation for the highest frequency component resolvable for a given size Airy disk, and by combining these we can show that the high frequency components in an image are indeed equivalent to (i.e., dependent on) rays passing through the outer periphery of the aperture (i.e., the size of the aperture).

Still, we are only showing an *equivalence* between the two—we are interpreting the Fourier transform in two different ways. The rays at the outer periphery of the aperture are *equivalent* to high frequency components of a 2-D Fourier transform. One is dependent on the other. The Fourier transform models the light modulation at the aperture.[8]

---

[8] You may not find it easy integrating all of the pieces we have discussed into this magnificent insight, but trust me, the mathematical exposition is no cake-walk either. As you know by now, these books are not *for dummies* books, but if you have made it this far you can surely handle this...and it's okay to go back and review the individual parts.

Wave Optics Mechanism

FOURIER DIFFRACTION

As a second example we have included a function in FFT15.00 to illustrate diffraction. This routine allows the user to specify a *slit width* (i.e., illumination pattern) which will transform nicely into the diffraction pattern observed with our laser and razor blades. Note that as the height of the slit illumination increases the height of the diffraction pattern decreases.[9] There's one special slit geometry you must try—a square. Select an 8 by 8 square slit and you will find that diffraction takes place in both the *x* and *y* axes (as you might have guessed). There are a couple of special examples included in this routine—if you use slit dimensions of 0,1 a hex-shaped aperture will be generated, and when you transform this function you will find a diffraction pattern displayed in *three* axes. If you select a slit of 1, 0 an octagon shaped aperture will be generated, the transform of which not only shows four axes of diffraction but also makes it apparent that, as the number of sides are increased, the diffraction pattern begins to approach a circle!

RESOLVING POWER

If a lens is placed behind the aperture it's reasonably apparent rays of a wavefront propagating in any direction will be focused into a single point, and the superposition of the rays must yield a diffraction pattern. Specifically, for rays parallel to the optical axis (i.e., at $\theta = 0°$) all of the increments along the wavefront will be *in-phase*, and the light moving in this direction will be unattenuated. As θ moves to either side of this forward direction, however, some of wavefront increments (as illustrated in Fig. 15.10) will not be in-phase, and therefore the total light intensity will be diminished. As the angle increases this attenuation will become greater and greater until we reach a null (as described on *p.* 177) when θ equals $\sin^{-1}(\lambda/D)$ where λ and D are as shown in Fig. 15.10. Note this phenomenon arises because of the Huygens *point source* model of radiation from a wavefront and the

Fig. 15.17 - Diffraction Pattern

---

[9]Those familiar with optics will note that we are also working with the amplitude function—not intensity.

summation of sinusoidal waves. Also note the similarity to the generation of a sinc function as shown in Table 11.2 (*p.* 131).

The bright spot formed by this diffraction phenomenon is called an *Airy disk*,[10] and determines the limiting resolution of any optical system. Note that the size of the diffraction pattern *and* the image are both proportional to the focal length of the lens; so, since *both* are proportional the focal length, we don't really gain anything (so far as resolving power is concerned) by increasing the focal length of the lens (i.e., the *magnification* is determined by the focal length, but once we reach a magnification that makes the Airy disk clearly visible, further magnification yields no new information). On the other hand, increasing the diameter of the lens (i.e., the aperture) reduces the size of the Airy disk, and is indeed the variable that controls resolution!

CONCLUDING REMARKS

We can't prove the equivalence of the wavefront at aperture plane and the 2-D transform via examples, but proving such things is not a major concern for us here. Proof is given in the advanced texts on Fourier analysis (and none of us will ever forget those). We *have* seen enough here to whet out appetite however. Understanding that the transform domain takes us back to the unfocused signal at the aperture causes the mind to race with the possibilities that might hinge on this mechanism. Can we take an *out-of-focus* picture, transform it back to the aperture, and then bring it back to the image domain in perfect focus? Can we, using the fractional frequency techniques we tinkered with in the *audio applications*, do equally amazing things with images? The answer is yes, of course, and in the next chapter we will explore the tip of the ice berg concerning how some of this is done...but just enough to help you decide....

---

[10]For Sir George Biddell Airy (1801-1892).

CHAPTER XVI

IMAGE ENHANCEMENT
CONVOLUTION/DECONVOLUTION

16.0 INTRODUCTION

Convolution may be loosely described as *filtering*. Now, filters have been around a long time (finding the *average* or *mean* is essentially a form of filter); still, most of us must advance to a certain level before we realize we don't quite understand this phenomenon.[1] Phenomena such as *filtering*, in both mechanical and electrical engineering, are ubiquitous today, leaving no doubt that, if we hope to understand this world we live in, it's imperative we understand this relationship known as *convolution*.

So, once again, let's ask, "What *is* convolution—what does it do?" We understand the meaning of multiplication, for example, as a simple attenuation or amplification of a signal. It's easy to understand multiplication of individual harmonics in the frequency domain; however, it's not immediately obvious what the effect will be in the time domain (especially when we multiply *all* the harmonics by some *function*). *The effect that occurs in one domain when multiplication is performed in the other is convolution. The unique results imply the ultimate sinusoidal nature of the signals multiplied, and the unique phenomenon of side-band generation when sinusoids are multiplied.*

In the time domain, attempts to understand "overshoot and ringing" (i.e., Gibbs phenomenon) in mechanical and electrical devices are nothing less than dismaying. Surely the simplicity of multiplication in the frequency domain is a much better explanation. The reality of frequency domain functions is something our ears have been telling us about since the invention of music—the reality of the frequency domain in optics is, today, just as obvious. Convolution just might be the most profound thing Fourier analysis has taught us in the last two centuries; so, let's see what we can do with this newfound knowledge.

---

[1] It's remarkable that many such phenomena were "invisible" only two centuries ago. Poisson was skeptical of Fresnel's wave theory of light and, by showing it predicted a bright spot at the center of a shadow, felt he had shown it to be incorrect (Arago saw this conclusion was a bit premature, and took the trouble to investigate).

16.1 OPTICAL TRANSFORM DOMAIN

Historically, imaging systems have provided an unending series of world class puzzles. Just as soon as one puzzle is explained, someone else points out even more puzzling phenomena, and the game is on again. An aperture placed at a lens (Fig. 16.1) will display its Fourier transform at the location of the image (if there is no *aperture proper*, the diameter of the lens will always provide one). So, if the lens is focusing an image at the location indicated in Fig. 16.1, we might (assuming the imaging process is bilateral) suppose the

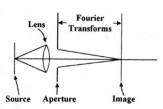

Fig.16.1 - Imaging System

transform of that image will be found at the aperture. This being so, it's apparent the transform of the image is multiplied by the aperture in the frequency domain...and this *frequency domain boundary* implies an *image domain* phenomenon very similar to the *Gibbs phenomenon* we know so well! In fact, it's the source of the diffraction pattern we call the Airy disk.

16.2 THE TRANSFER FUNCTION

Developing the notion of a transfer function will be helpful. As is frequently the case, this *advanced concept* is really the simplest imaginable relationship. It is the frequency domain output of a *system* divided by its frequency domain input. If we drive a linear system with a signal defined as:

$$A(f) = A\ Cos(2\pi ft)\quad [\text{for all } f]$$

then, as we know, the output will be a sinusoid of the same frequency (modified only in amplitude and phase):

$$B(f) = B\ Cos(2\pi ft + \phi)$$

The transfer function, then, is defined as:

$$T(f) = \frac{B(f)}{A(f)} \qquad (16.1)$$

[The (*f*) in these equations, as usual, implies *all* frequencies in the signal. An impulse function serves nicely for this evaluation, of course.]

Image Enhancement

Conceptually, this notion includes anything and everything that linear systems might do to signals that pass through.[2] This is just a very general way of saying we want to talk about *whatever* a system might do to a signal (e.g., low pass filter, amplify, differentiate, etc.).

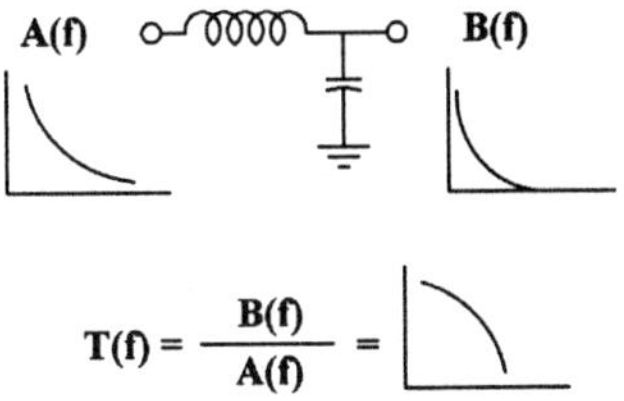

$T(f) = \dfrac{B(f)}{A(f)}$

Fig. 16.2 - Xfer Function

So, if we know the input function, and the transfer function, the output is then given by:

$$B(f) = A(f)T(f) \quad (16.2)$$

This multiplication in the frequency domain is *convolution* in the space/time domain—it's filtering—and to see the time domain signal we must inverse transform this result.

The other beneficial use of transfer functions is this: when we know the *output function*, and either *know* the transfer function, or can reasonably approximate its characteristics, we can find the original input:

$$\dfrac{B(f)}{T(f)} = A(f) \quad (16.3)$$

This simple relationship describes *de-convolution*—our efforts to recover high fidelity sound from old recordings for example. Something this simple *must* work...let's look at a two dimensional example.

16.3 CONVOLUTION AND IMAGE RESOLUTION

The transfer function includes those things that *modify* the input signal. If we have a *perfect* lens, for example, the transfer function will include only the aperture function we just described. If, however, you're using a cheap camera with a mass-produced, untested plastic lens—or you get mud on the lens of the expensive camera you borrowed from a rich friend—the transfer function will include a great deal more. In any case the transfer function will *multiply* the image transform (Fig. 16.1), and the image will be the *convolution* of this transfer function with the object. So,

---

[2] We understand, of course, that when we limit our discussion to linear systems we disallow things like backlash, pump valves, diodes, logarithmic converters, etc.

just as the aperture creates the Airy disk (yielding a less than perfect image), lens aberrations degrade the result. When we deal with this sort of problem it's a great simplification to think in terms of *multiplication* of transforms in the frequency domain. Clearly, if the distortion is caused by some action in the aperture plane (e.g., mud on the lens), the distorting function will be convolved with the image and becomes part of the transfer function. To remove this sort of distortion we must *divide* the frequency domain function by the *transfer function*! In fact, if we knew the *exact* equation for the transfer function we could *exactly* reproduce the input! That's what Eqn. 16.3 says! *[Unfortunately, you can't divide by zero! The aperture function multiplies everything outside the aperture by zero so there's no way to get that portion of the input back...(sigh).]*

16.4 BRINGING CONVOLUTION INTO FOCUS

Okay, let's look at how, exactly, the above procedure works. If the picture of your girlfriend "came out" you will get something like Fig. 16.3; but, suppose things didn't come out this way! Suppose you *really did* get mud on the plastic lens. *[Note: the program for this chapter, FFT16.00, will actually perform the steps we describe in the following.]*

Fig. 16.3 - Girlfriend

We need to make an adjustment—Figure 16.3 was cropped from a much larger photograph (we needed a 64 x 64 pixel array for this illustration) and, as you can see, we have a case of the "jaggies." We can make significant improvements by oversampling and filtering (see Fig. 16.4). [Routines to do this are provided in program FFT16.00].

Fig. 16.4 - Improved

Okay, we will simulate deterioration of the image by first generating a Gaussian function (using a function diameter of 2.0). Transform this Gaussian function and save the transform (under the *Modify Transform* subroutine). Next, load the picture of your choice (our software is limited so it will have to be a gray scale, *.BMP format). Transform the picture and then *multiply* the transform with

Image Enhancement

the stored function (again, under the *Modify Transform* option). Next we reconstruct the image and save it to the disk under some convenient filename (you may look at this fuzzy picture later, but you can see it now in Fig. 16.5). Okay, without changing anything, transform this degraded graphic back to the frequency domain. Under the *Modify Transform* menu there's an option to *Divide Transform with Stored Function*. The *stored*

Fig. 16.5 - Fuzzy

function, of course, is the same one we just used to multiply this transform. If we *divide* by this function, we should get the same thing we had originally...what could go wrong?

Fig. 16.6 - Recovered

If, in fact, you divide by this same Gaussian function, you *will* (pretty much) get the original picture back...but check the deconvolved spectrum carefully—it's "growing hair" at the highest frequencies. You can play barber and cut this stuff by using the *Truncate Spectrum* option under the *Modify Transform* menu...but we should be *very concerned* about things growing hair...let's look closer.

Start all over again, but this time make the Gaussian function diameter = 4.0. Go through the same process (this time there's no need to save the picture—we've already done that—it's just another fuzzy picture). The thing we're interested in here is the final deconvolved frequency domain function and...WOW! This is pretty bad! This stuff doesn't need a barber—it needs a hay-mower!

We better talk about this. What we have just done is simply multiply a bunch of numbers by a bunch of numbers...and then *divide* the same bunch of numbers (by the same numbers) and, mathematically, we should be back where we started. Nonetheless, we do *not* have the same thing we started with! It would appear our computer is not very good at arithmetic. =● It's okay, all it takes to correct this is one line of code:

```
10 DEFDBL A-Z  ' THIS IS ALL IT TAKES
```

That's all it takes...(well, you *will* also have to change the Dimension statement, etc., etc., etc...). Run the identical illustration as above under double precision and it will work.

The reason it takes double precision to get good arithmetic in the above illustration is that a Gaussian function greatly attenuates the higher frequencies (Fig 16.7). We are multiplying the high frequencies of our transform by very small numbers and then performing a multitude of multiplications and *additions* on these numbers (i.e., the FFT operations). When we later divide the transform by these same small numbers we find the *original data* has been lost in noise.[3] The point is this: unless we're careful, our attempts to enhance the image may well *screw-up* the data so badly things become unrecoverable (if they weren't hopeless to begin with). If you drive the numbers into the noise with single precision arithmetic, you yourself may make it impossible to recover the data.

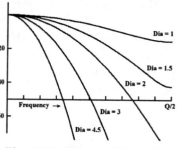

Fig. 16.7 - Xform of Gaussian

Okay, except for the *noise* thing, what we have just done is pretty simple. We have multiplied two functions (the image transform × a hypothetical transfer function) and then divided them. It would have been surprising *only if it hadn't* worked; but, some of you may suspect enhancing *real* fuzzy pictures will be more difficult. For example, if we get mud on our friend's camera, we won't *really* know the transfer function! *Real* fuzzy images just might be a little tougher!

16.5 FINDING TRANSFER FUNCTIONS

Okay, let's see what we can do with this de-convolution technique. It should be apparent that transfer functions exist which can convert any picture into any other picture (e.g., that we *can* convert an out-of-focus picture into an in-focus picture). Furthermore, we could always find this transfer function by dividing the transform of the desired picture (i.e., the in-focus picture) by the transform of the picture to be converted (i.e., the out-of-focus one).

---

[3] We can multiply by small numbers and things work well enough, but when we add very small numbers to very large ones the least significant bits of the small ones (or even the MSBs) may be truncated. You may recall our discussion back on *p.* 137.

Image Enhancement

It's obvious, of course, that if we have an in-focus version of the picture, we probably don't need to go through this exercise; still, there *is* a reason to conduct this experiment. If we deliberately set up a camera, take a picture of a "fixed" object, then de-focus the camera and take a second picture, we will have the raw material to determine the necessary de-convolving function when we *do* get an out of focus picture!...Aha![4]

This, in fact, is precisely what we set-out to do. We allowed that, if we find the *transfer function* resulting in a less than perfect result, we could then obtain the original input by dividing by this transfer function! We should obtain the required deconvolution function if we divide the transform of the out-of-focus picture by the in-focus one. Now, as we know, we will pick up truncation and rounding *noise* in this process; nonetheless, we should get a useful replica of the *function* required to bring pictures into focus.(?) If you have access to a CCD camera, VCR camera with digital converter, or a conventional 35mm camera and scanner, you can try this project, but you may find it more "enlightening" than you wanted— working with real data is always more frustrating than working with idealized experiments. You will find it desirable to start with a simple *target* (e.g., small black circle on a white background or vice-versa—see comment at the bottom of *p.*186). You might find that your key-chain laser (made into a point source by projecting it through a lens) is quite useful here.

We should point out that this technique of deconvolution is almost too general. If, for example, you divided the transform of Mickey Mouse by the transform of someone else, you could make, say...your favorite politician...look like Mickey Mouse (that would be a trivial case, of course).[5] In other words, not only can we make an *out-of-focus* picture *in-focus*, we can pretty well turn anything into anything (such is the power of this sort of *filtering*). What we really need are well-defined functions for the distortion we are trying to correct...ones we can apply in an iterative process, measuring whether each step makes things better or worse...and reacting accordingly. Surely Mother Nature is doing something simple when she creates out-of-focus pictures.(?)

There are other mechanisms that degrade the image, and here deconvolution may be very helpful—let's look at this.

---

[4] And if it doesn't work we will at least be thereby *enlightened!*
[5] No offense intended Mickey.

16.6 ENHANCING RESOLUTION

The picture of Fig. 16.8 was scanned from a small segment of an 8" by 10" photograph into the .BMP format. (*NOTE: All the pictures in this chapter are handled in the *.BMP graphics format. It's one of the easiest formats to understand (which is why we chose it) and it's the only format provided for in program FFT16.00.*) [Note: while the half-tones presented here may leave something to be desired, the relative (before and after) quality of the photos is a fair representation of what was achieved.]

Figure 16.8 - Moon Shot

In enhancing this photograph we have done little more than improve the contour of the frequency domain spectrum, using the same technique we have been describing. We used a Gaussian function (as the deconvolving function), for the same reasons we used it in the preceding illustration—it's a good approximation to what actually happens in less than perfect optical systems. We might have done better had we taken a picture of a star (under the identical conditions as this picture was taken) and used its transform as the deconvolving function. If you choose to experiment with that technique, remember the deconvolving function must be provided to considerably better resolution and precision than the picture you hope to improve—that's one of the major advantages to using a Gaussian or other purely mathematical function.

In Fig.16.9 we begin to see the good news and the bad news. We *can* bring up detail, but with this 8-bit data format, we quickly begin to bring up the digital noise. We have probably harped on this enough, but things can be greatly improved by using more bits of data, and by *integrating* (i.e., summing together) a large number of images to drag the detail out of the noise. Note: this was originally in color (we converted to a gray-scale

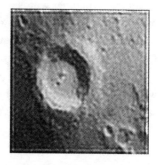

Fig. 16.9 - Enhanced Moon

Image Enhancement

before enhancing)—we could have probably done a little better by enhancing the three colors separately and combining them afterwards.

If you have access to a CCD camera mounted on a telescope, and a modern multi-media computer, you can go pretty far using an empirical approach. It's a rewarding experience when you see real improvements appear after you have worked on an image for a period of time. This really is *good stuff*; however, you should always keep in mind that there's an ultimate limit to the resolution attainable with any given telescope objective. The image is indeed convolved with the system aperture, and any harmonic components that lie beyond the aperture function are zeroed-out...and lost forever.[6]

16.7 DATA LENGTH PROBLEMS

Okay, now that we understand the basic idea of deconvolution let's look at some of the things that affect image enhancement in practice. A close relative to the double-precision bomb is the data precision bomb. Most of the popular graphic file formats (*.JPG, *.BMP, etc.) only employ eight bits of *resolution/dynamic range*; so, if 7 *decimal* digits blow up, how do 8 *binary* digits fare? If we repeat the *girl-friend* illustration but, after reconstructing the *fuzzy* image, truncate the gray-scale to 8 bits, we will have the thing we usually get when we scan an imperfect picture.[7] Now, as we know, the spectrum of a Gaussian function *rolls-off* with frequency (Fig. 16.7), and if we multiply the harmonics of a picture by this function, we eventually attenuate those harmonics below the least significant bit of our A/D (about -50 db for 8 bits). In our *convolution/deconvolution* illustration we multiplied the 8 bit resolution pictures with *real numbers* (7 decimal digits); so, when we *enhanced* this fuzzy image we pretty much got our original 8 bit data back. When working with *real* fuzzy images, however, the Gaussian function we're concerned with isn't the mathematical function we used in the preceding example, but the quasi-Gaussian function caused by mud on the lens of your camera, or etc. In this case the high frequency components have long since been lost by

---

[6]There is, as we noted in our study of audio, the technique of guessing what the image should look like and then minimizing the error between the guess and actual image. If we really do know what the image will look like this can work pretty well.

[7]There's an option from the *Main Menu* of FFT16.00 which is called *Change Gray-Scale*. It allows setting the data in the array back to an integer number of data bits.

our 8-bit A/D, and dividing the transform of this 8 bit data by our mathematical Gaussian transform can't bring back the lost information—it will only create random numbers in the high frequencies—noise.

The message is, in the case of data bits, more is better, but 16 bits only give 96 db of dynamic range...and from Fig. 16.7 it's apparent that seriously degraded images will need more than that. *[Note: the maximum resolution frequency is, of course, the Nyquist (indicated as Q/2 in Fig. 16.7). Our efforts here only concern harmonics below the Nyquist.]* You will get a better feel for this as you work with it.

You can experiment with the effect of data length by converting the degraded image to various data length words (e.g., 8 - 16 bit resolution) and then try to deconvolve it in the usual way. You will, of course, get the overgrown "hay" at the high frequencies (you can try this in double precision if you like, but it's pretty obvious that the problem is data word length). In Fig. 16.10 we compare attempts to recover images

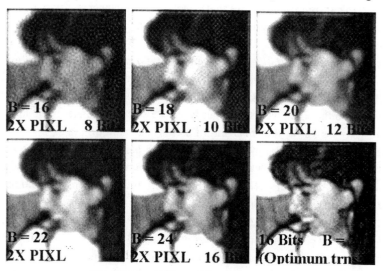

Fig.16.10 - Effect of Data Resolution

with increasing numbers of data bits. Now, this *isn't supposed* to represent the best you can do—the *best you can do* is as much art as science, and appearance is greatly influenced by things like *brightness* and *contrast,* which have little to do with our struggles within the FFT. Some people can get much better results than others—the pictures of Fig. 16.8

Image Enhancement

were obtained with a relatively *fixed* routine, keeping things the same for each picture (a not completely successful effort, obviously). The last picture (bottom right) was obtained after an unjustifiably long effort (as I said, some are better at this than others). Compare these to Fig. 16.4 which is the best we can do with a perfect recovery.

16.8 ABOUT THE NOISE "PROBLEM"

So, if 16 bits only provides 96 db of dynamic range (software is moving toward 48 bit graphics with three 16 bit colors—maybe whining works after all) can we really do anything significant to enhance images that are seriously degraded? We really need more data bits, and in that regard we need to consider another problem.

The "grass" we encountered in the *fuzzy image* illustrations is *digital noise*—we run into truncation and rounding errors which affect the data in a more or less random manner. Increasing the number of bits in our A/D converter will alleviate this problem; however, as we continue to increase the number of bits, we will eventually run into the *system* or *electro-optic* noise. Electronic noise is also random stuff which can't be corrected, leaving us impaled on the horns of a dilemma: limiting the number of data bits creates digital noise, but increasing the number of bits brings in electro-optic noise! This sounds like *Catch-22*.

Actually, we should *not* be upset if the least significant bits of our A/D converter are displaying noise! In fact, seeing noise in these least significant bits should warm the cockles of our hearts, for this is the solution to our dilemma—this allows us to *go beyond* the 8-bit resolution of the A/D converter! All we have to do is *sum* these noisy pictures into an *accumulator*, and as we continue to add-in noisy frames of data the *signal* will increase linearly...but not the noise...*noise is a random phenomenon*. If we sum a series of random numbers, the *summation* will stagger off in a *random walk*. We can show that the RMS value of this random walk function will only increase as the *square root* of the number of added frames. So, as we add noisy picture on top of noisy picture, the *signal* increases linearly, the noise increases as $\sqrt{N}$ (N = number frames), and the *signal to noise ratio* will improve as $\sqrt{N}$. If we do this correctly the resolution will also improve (i.e., the number of bits in the data word will increase). [There's an illustration of this in Appendix 16.2.]

16.9 A FINAL WORD

There are many other applications of the 2-D FFT, many of which are more interesting than our chosen example; however, given the limitations of this book it seemed best to consider the phenomenon of convolution/deconvolution. Now, as we noted in the last chapter of *Understanding the FFT*, there's a way to perform convolution without involving the Fourier transform—the *convolution integral*. As we know, this involves multiplying and integrating the functions, then shifting and multiplying and integrating again, etc., etc., etc.—a convoluted process that could be appealing *only* to a mathematician. It's difficult to imagine that Mother Nature would abuse us with physically real implementations of such a process; still, since multiplication in the frequency domain *is* convolution in the time domain, it's not a matter of one viewpoint being the "right" point of view—both describe the same thing....

So far as physical reality is concerned, the understanding of convolution as multiplication in the frequency domain seems much more likely...*if* you accept the frequency domain harmonics as physically *real*. Now, when we deal with light waves this is no great stretch since we are obviously dealing with sinusoidal waves...and when we deal with *sound* we're likewise dealing with sinusoids. In such instances there can be little doubt that multiplication in the frequency domain is the *real* point of view. On the other hand, there are a multitude of instances when it's not obvious that sinusoids are involved at all (e.g., linear motions of objects, described by Sir Isaac's famous F = Ma)? We might argue, as we noted in Chapter 2, that the *real world* includes both phenomena of differentiation and integration. That is, *inertial acceleration*s co-exist with *spring displacements*—objects thrown into the air are pulled back by the "spring" force of gravity.[8] Sturm and Liouville have patiently explained that systems described by second-order differential equations tend to *oscillate*—implying underlying sinusoidal mechanisms—and even when the oscillations are damped-out completely the underlying sinusoidal mechanisms are still valid. Nonetheless, as we noted earlier, phenomena following the e^x *natural growth law* do not qualify for Sturm and Liouville's blessings.

---

[8] The physics of electricity and magnetism make this point strongly as the ubiquitous nature of *stray* capacitance and inductance is grasped.

Image Enhancement

As our understanding of the world about us progresses our viewpoints change. Newton saw the phenomenon of light cleanly and clearly as a rain of particles; nonetheless, people like Huygens, Young, Fresnel and Fraunhofer recognized that phenomena such as interference and diffraction dictated that light be viewed as waves. Eventually it became impossible to deny the wave-like nature of light, and the wave theory won out over the particle theory (unfortunately, and all too soon, it later became impossible to deny the particle nature of light, thereby confounding everyone on all sides). Even worse, it shortly became apparent that basic particles of matter (which, by this time, *everyone* agreed had a particle nature) also exhibited a wave-like nature...it's enough to try your patience!

What Fourier analysis is really telling us is that both these observations—both of these points of view—are the same thing (at least when we're dealing with second order systems). If the summation of sinusoids can yield a replica of any waveshape we can measure, then surely we can decompose any measurable waveshape into sinusoids. Depending on how we manipulate a signal in its two domains, we can make either the wave nature or the particle nature apparent, as we have tried to illustrate in this book. In any case, there's nothing supernatural nor magical going on here, and as we understand this, we begin to understand Fourier analysis.[9]

---

[9]Fourier analysis can be extended to the third dimension. The way to do this is more-or-less apparent from our discussion of the 2-D transform. The implication is that 3-D objects are a superposition of sinusoidal waves...talk about interesting puzzles. It has occurred to me that a third book on the 3-D transform might be interesting...but so far I have been successful in putting it out of my mind.

APPENDIX 1.1

AVERAGE VALUE

The Pythagoreans, some 500 years before the birth of Christ, defined three different *means* which they termed "arithmetic, geometric, and harmonic." (Actually they defined ten different proportions they called "means" but we won't go into *that*.) It's the notion they called *arithmetic mean* that interests us here—that's what we refer to as the *average* value.

The arithmetic mean is the point midway between any two arbitrary numbers. The number 5 is the average value of 3 and 7—it's an equal *distance* from 5 to either 3 or 7—that's what we mean by the term *arithmetic mean* (or *average*). Let's look at this notion arithmetically:

$$(Av - n_1) = (n_2 - Av) \quad \text{----------} \quad (A1.1)$$

where n_1 is the smaller number and n_2 is the larger [e.g., $(5-3) = (7-5)$]. If we solve (A1.1) for Av we find:

$$Av = (n_1 + n_2) / 2 \quad \text{-----------} \quad (A1.2)$$

The notion of a *mean* extends to more than two numbers, of course. The number 5 is also the average of 3, 4, 6 and 7. It's the *middle point* of this set of numbers. If you expand this exercise, you will eventually conclude that the average of any set of N numbers is:

$$Av = 1/N \sum n_i \; (\textit{for } i = 1 \textit{ to } N) \quad \text{-----} \quad (A1.3)$$

That is, we add up the numbers and divide by the number of data points.

Now, we may use this notion to find the average value of a *function* over some range. For example, we would solve the very simple function $y = \text{INT}(x)$ for its average value from zero to 4 as follows:

$$(0 + 1 + 2 + 3 + 4)/5 = 10/5 = 2$$

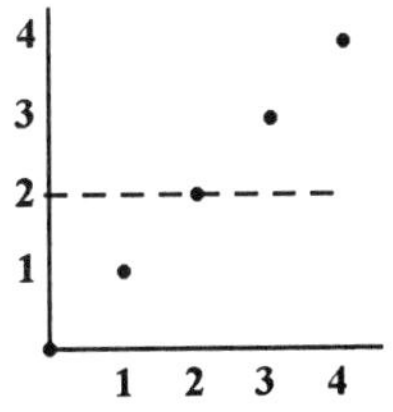

Fig. A1.1   The average value of this linear function between 0 and 4 is 2—we included zero—the average value from 1 to 4 is 2.5!

Well and good, but suppose, for example, we want to know the average value of the function $y = \text{INT}(x^2)$? The above technique yields:

$$(0 + 1 + 4 + 9 + 16)/5 = 30/5 = 6$$

but from the calculus the mean value (designated as $\overline{y}$) is:

$$\overline{y} = x^{(n+1)}/(n+1)x = x^3/3x = 4^3/(3\cdot 4) = 64/12 = 5.333....$$

We have a little discrepancy here—which answer do we really want?

Our question is: "Why is the *mean value*, as determined by the calculus, not the same thing as the *arithmetic mean*, as determined by summing the data points and dividing by N? The following program may help shed a little light on this matter:

```
' ***************************************************************
' * PROGRAM TO COMPARE AVERAGE VALUES OF CURVES OF THE FORM     *
' * (Y = X^N) AS OBTAINED BY:  1) THE MEAN VALUE THEOREM        *
' * 2) SUMMATION DIVIDED BY N  3) THE TRAPEZOIDAL RULE          *
' ***************************************************************
CLS : DEFDBL A-Z
INPUT " POWER TO RAISE INDEPENDENT VARIABLE TO IS "; N
INPUT " CALCULATE OVER DOMAIN (X MAX.-INTEGER)"; XMAX
10 INPUT " USE A STEP SIZE OF: (E.G., 0.5)"; SZ
PRINT : PRINT
PRINT "MEAN VALUE DETERMINED BY CALCULUS IS "; (XMAX^(N+1)/(N+1))/ XMAX
PRINT
' **********************************************************
YT = 0 ' INITIALIZE ACCUMULATOR
FOR I = 0 TO XMAX STEP SZ ' SOLVE ARITHMETIC MEAN
YT = YT + I ^ N
NEXT I
YAV = YT / (XMAX / SZ + 1)' ADD 1 TO COUNT THE "ZERO" DATA POINT
PRINT "THE ARITHMETIC MEAN IS "; YAV: PRINT
YT = YT - (XMAX ^ N / 2): YAV = SZ * YT / XMAX
' **********************************************************
PRINT "THE AVERAGE DETERMINED BY THE TRAPEZOIDAL RULE IS "; YAV
PRINT "MORE (Y/N)";
90 A$ = INKEY$: IF A$ = "" THEN 90
IF A$ = "Y" OR A$ = "y" THEN 10
END
```

If you experiment with this program you will find that as the number of data points is increased, the arithmetic *average* approaches the *mean* as determined by calculus. Calculus finds the mean of an *infinite number of data points*, and as we take more points, the arithmetic mean approaches the same value. The calculus finds the *mean value* of a continuous function—the *arithmetic mean* gives the average of a finite number of data points. If you're working with a continuous function you must use the calculus—the average of a finite set of numbers is given by Eqn. (A1.3).

APPENDIX 4.1
DIFFERENTIATION AND INTEGRATION VIA TRANSFORMS

A4.1 DIFFERENTIATION

Let's make two fundamental points:
1) In Chapter 2 we noted the harmonic components could be *modified* but, in a linear system, the resulting function always remained a summation of these *same frequency* sinusoids.
2) In general, this function is expressed as:

$$f(t) = A_0 + A_1\text{Cos}(t) + B_1\text{Sin}(t) + A_2\text{Cos}(2t) + B_2\text{Sin}(2t) + ...) \quad (A4.1)$$

Now, we may differentiate (or integrate) a polynomial by differentiating (or integrating) the polynomial term by term; so, we may differentiate the Fourier series of Eqn. (A4.1) by differentiating the harmonics term by term. In the DFT this turns out to be an incredibly simple operation that requires nothing we would really call calculus. Let's look at this.

As noted in Chapter 2, the derivative of a sine wave is a cosine wave (p. 10-12), and if a cosine wave is just a sine wave displaced by 90°, it's apparent the derivative of a cosine wave will be a negative sine wave, etc. That is, the derivative of $\text{Sin}(x) = \text{Cos}(x)$ and the derivative of $\text{Cos}(x) = -\text{Sin}(x)$ [i.e., $d\text{Cos}(x)/dx = -\text{Sin}(x)$]; however, when we deal with *time-based* sinusoids (i.e., $\text{Sin}(\omega t)$, the derivatives are multiplied by ω (i.e., $d\text{Cos}(\omega t)/dt = -\omega \text{Sin}(\omega t)$, where $\omega = 2\pi f$). *(Those unfamiliar with the calculus will have to take my word for this.)*

The harmonics are designated as complex phasors, composed of a cosine and sine term. So, to find the derivative of a time-based sinusoid, then, we need only multiply all its phasors by $2\pi f$ (i.e., ω), and shift the phase by 90°. Mathematically, we convert the frequency domain function to its derivative by simply *multiplying* each harmonic by the complex variable $0 + i2\pi f$ (where $i = \sqrt{-1}$ and f would be the harmonic number).

Okay, but for some of us there's a deep, dark chasm between the *explanation* (such as given above) and actually finding derivatives and integrals using an FFT computer program. It will be a worthwhile exercise to modify the basic FFT program we developed in *Understanding the FFT* to differentiate and integrate functions (see Ch. 6, this book, pp. 51-56. A basic differentiation routine might look something like this:

```
1050 CLS: PRINT SPC(30); "CALCULUS MENU": PRINT : PRINT
1060 PRINT SPC(5); "1 = DIFFERENTIATE FUNCTION": PRINT
1062 PRINT SPC(5); "2 = INTEGRATE FUNCTION": PRINT
1066 PRINT SPC(5); "9 = EXIT": PRINT
1070 PRINT SPC(10); "MAKE SELECTION :";
```

```
1080 A$ = INKEY$: IF A$ = "" THEN 1080
1090 A = VAL(A$): ON A GOSUB 1100, 1200
1099 RETURN
1100 REM *** DIFFERENTIATE FUNCTION ***
1110 FOR I = 0 TO Q3: KDF = I * P2 ' I = HARMONIC, PI = 3.1416...
1120 CTemp = -KDF * S(I): S(I) = KDF * C(I): C(I) = CTemp
1130 CTemp = KDF*S(Q-I): S(Q-I) = -KDF*C(Q-I): C(Q-I) = CTemp
1140 NEXT I
1142 C(Q2) = 0: S(Q2) = 0
1150 NM$ = "DERIVATIVE": RETURN
```

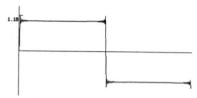

Fig. A4.1 - Square Wave Function

We didn't really employ the formality of complex multiplication here. Since sine components obviously become cosine components and cosines become negative sines, we simply multiply and replace terms. This is a *powerful* tool—we can find derivatives even when we know virtually nothing about the function (nor the calculus = ●).

If you type in the generic FFT program of Chapter 6, *p*.51, along with the differentiation routine, you can experiment with this technique to your heart's content; however, there *are* idiosyncrasies and pitfalls—so we should talk about these.

Fig. A4.2 - Derivative of Square Wave

A4.2 THE TIME SCALE FACTOR

So far, in our work with the FFT, we have assumed a time *domain* of unity (i.e., one second). That, of course, will seldom be the case; but, as far as the FFT is concerned it makes no difference. We will always fit exactly one sinusoid into the domain for the fundamental, and the harmonics will always be integer multiples of this fundamental. If, for example, we study the offset voltage drift of an operational amplifier, and take one sample per day for 1024 days, the harmonics will be in units of N cycles/1024 days. The derivative of this data will be expressed in volts per 1024 days. [If we should need the drift rate per day we simply divide the data by 1024].

Having noted this, it's apparent the reference period for our sinusoids (i.e., T_0) is always equal to the time of the total sample interval, and that the time between samples (Δt) must be T_0/Q (where Q is the total

Differentiation and Integration

number of samples). Formally, the argument for our sinusoids will be $2\pi n \Delta t / T_0$. If we sample a 1 Hz triangle wave (amplitude of -1 to +1 volts) at 512 samples/second for one second, it's more or less obvious the derivative will be a square wave of ± 4 volts/sec. (that is, the signal ramps up from -1 to +1 in 1/2 second which is a rate of 4 volts/second). How does this work out in our FFT differentiation program? Do we get quantitatively correct results? To check this we need to modify the generate function routine of our program to the following:

```
600 ' *** TRIANGLE FUNCTION ***
.
.
610 FOR I = 0 TO Q - 1: C(0, I) = 0: S(0, I) = 0
612 FOR J = 1 TO Q2 - 1 STEP 2 ' SUM ODD HARMONICS
620 C(I) = C(I) + (SIN(K1 * I * J) / (J * J))
630 NEXT J' SUM IN NEXT HARMONIC'S CONTRIBUTION
640 C(I) = C(I) * 8 / PI^2 ' CORRECT FOR UNIT AMPLITUDE
650 NEXT I ' CALCULATE NEXT DATA POINT
```

If you try this you will find it works fine, but suppose our actual digitized function was a 100 Hz triangle wave of amplitude ± 1 volt? In that case (assuming we still digitize one cycle in 512 samples) the total data interval will be 0.01 second, but the data will be identical to the 1 Hz waveform. The derivative will still yield ± 4 but we must divide by the 0.01 seconds and the correct derivative will then be 400 volts/sec. *While we do not carry the time scale factor into the DFT, we should keep it safe somewhere so that we can calculate and display the correct amplitude/time/frequency data for the end user.*

A4.3 INTEGRATION

It's reasonably apparent that, if we can differentiate a frequency domain by multiplication, integration may be accomplished by division. We may find the integral of a Fourier series by simply *dividing* by the same complex variable we used to differentiate (i.e., $0 + i2\pi f$—*surely this will get us back to the original function..no?*). A preliminary integration routine might look something like this:

```
1200 REM *** INTEGRATE FUNCTION ***
1210 FOR I = 1 TO Q3: KDF = I*P2 ' I = HARMONIC, P2 = 2*PI
1220 CTemp = S( I) / KDF: S( I) = -C( I)/KDF:C(I)= CTemp
1230 CTemp =-S(Q-I)/KDF:S(Q-I) =C(Q-I)/KDF:C(Q-I)=CTemp
1240 NEXT I
1242 C(Q2) = 0:C(0) = 0
```

Let's use this *preliminary* routine as given here to explore the potential of this technique, and then we will talk about its shortcomings.

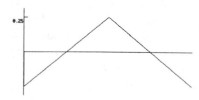

Fig. A4.3 - Integral of Square Wave

Suppose we begin by integrating the square wave. For this we must first create a square wave, then modify the spectrum (i.e., execute the *integration subroutine*) and finally perform an inverse transform to see the results of our handiwork (see Fig A4.3).

You may do repeated integrations (or differentiations) of functions; so, having found that the integral of a square wave is a triangle wave, we may either proceed to the integral of a triangle; or, we may differentiate this function just to verify that we get our square wave back.

There's a particularly interesting exercise you can do here—integrate the square wave more than once, then differentiate it repeatedly to see if you get your square wave back. You will need to extend this exercise to about four integrations/differentiations at least. If you are experienced with this sort of operation the results may not surprise you—but maybe they will. Also try this using double precision arithmetic.

Now, as we hinted above, there's a glitch with the routine given for integration here. If we attempt to divide the constant term (i.e., the zero frequency component) by *zero*, we will be doing something stupid. The computer will stop, of course, and tell us we have tried to perform an illegal operation—that division by zero is not allowed. So, in the above routine, we have simply *zeroed-out* the constant term...=●

Note: There's a way around this little problem. The trick is recognize that a constant offset would be identical to the first alternation of a square wave if the array were doubled in size (we'll talk more about this in Ch. 9). We will not go into this further but leave it as an extra credit exercise for the really <u>intelligent</u> readers...the rest of you probably couldn't handle it anyway....!^)

APPENDIX 6.1

TRANSIENT ANALYSIS PROGRAM

The two programs for Chapter 5 and 6 have been written so that they can be *merged* together with a minimum of headaches. First you will need to load the Mesh Analysis program ACNET05.BAS (*p.* 37), delete everything up to line 10000, and then save what's left under some temporary name (e.g., MESHTMP.BAS). Note that, for QuickBasic, you will have to save this as an ASCII file so that it can be merged later. Next load the Gibb's Phenomenon (FFT06-01) program from Chapter 6 (*p.* 51) and immediately save it as TRANALYS.BAS (or some name you prefer). Then *merge* MESHTMP.BAS to the end of this new program.

Okay, now we will have to make some modifications:

```
20 DIM C(Q), S(Q), KC(Q), KS(Q), COMP(3,KB,KB), Z(2, KB, KB+1)
.
1066 PRINT SPC(5); "4 = CIRCUIT CONVOLUTION": PRINT
1067 PRINT SPC(5); "5 = INPUT CIRCUIT": PRINT
.
1082 A = VAL(A$): ON A GOSUB 1100, 1200, 1300, 1400, 20000, 1990

1400 REM * CONVOLVE WITH CIRCUIT XFER FUNCTION *
1402 CLS : PRINT : PRINT
1404 PRINT SPC(10); "CONVOLVING XFER FUNCTION WITH STEP FUNCTION"
1410 FOR I3 = 0 TO Q3
1412 GOSUB 10000' COMPUTE XFER FUNCTION
1424 CAT = (ATTNR * C(I3))-(ATTNI * S(I3))' COMPLEX CONVOLUTION
1425 SAT = (ATTNI * C(I3))+(ATTNR * S(I3))'IMAGINARY TERM
1426 C(I3) = CAT: S(I3) = SAT' REAL FREQUENCIES
1427 C(Q-I3) = CAT: S(Q-I3) = -SAT' NEGATIVE FREQUENCIES
1428 NEXT I3
1430 I3 = Q2: GOSUB 10000
1434 C(Q2) = (ATTNR * C(Q2)):S(Q2) = 0
1494 RTFLG = 1: RETURN

10000 F = I3: FREQ = I3
10010 GOSUB 11000' CALC IMPEDANCE MATRIX
10020 GOSUB 12000' SOLVE FOR CURRENTS
10030 ATTNR = Z(1,KB,KB+1)*RTERM:ATTNI = Z(2,KB,KB+1)*RTERM
10090 RETURN

20002 PRINT: INPUT "ENTER NUMBER OF MESHES"; KB
20004 REDIM COMP(3, KB, KB), Z(2, KB, KB+1)
```

You should also substitute the generate square wave function shown on page 65, but that's all there is to it! You can now get the transient response of passive linear networks. Try it.

APPENDIX 7.1

POSITIVE FREQUENCY ONLY STRETCHING

This program is described in Appendix 5.1 of the companion volume *Understanding the FFT*. The only difference is *this* DFT only extracts the *positive frequencies*. This is easily and simply accomplished in line 206 by making the loop step from 0 to Q/2 (we must include the Nyquist). You should look at the Transform/Reconstruct routine—if for no other reason, just to remind yourself of the simplicity and beauty of that routine we call the DFT.

```
2 REM  *********************************************
3 REM  ** (DFT07.01) GENERATE/ANALYZE WAVEFORM  **
8 REM  *********************************************
10 Q = 32: PI = 3.141592653589793#: P2 = 2*PI: K1=P2/Q: K2=1/PI
14 DIM C(2, Q), S(2, Q), KC(2, Q), KS(2, Q)
16 CLS : FOR J=0 TO Q: FOR I=1 TO 2: C(I,J)=0:S(I,J)=0: NEXT: NEXT
20 CLS : REM *    MAIN MENU    *
22 PRINT : PRINT : PRINT "          MAIN MENU": PRINT
24 PRINT " 1 = STRETCHING THEOREM": PRINT
31 PRINT " 2 = EXIT": PRINT : PRINT
32 PRINT SPC(10); "MAKE SELECTION";
34 A$ = INKEY$: IF A$ = "" THEN 34
36 A = VAL(A$): ON A GOSUB 300, 1000
38 GOTO 20
40 CLS : N = 1: M = 2: K5 = Q2: K6 = -1: GOSUB 108
42 FOR J = 0 TO Q: C(2, J) = 0: S(2, J) = 0: NEXT
44 GOSUB 200: REM - PERFORM DFT
46 GOSUB 140: REM - PRINT OUT FINAL VALUES
48 PRINT : INPUT "C/R TO CONTINUE"; A$
50 RETURN
100 REM  *********************************************
102 REM  *           PROGRAM SUBROUTINES           *
104 REM  *********************************************
106 REM  *        PRINT COLUMN HEADINGS           *
108 PRINT : PRINT : IF COR$ = "P" THEN 116
110 PRINT "FREQ     F(COS)       F(SIN)      FREQ     F(COS)       F(SIN)"
112 PRINT
114 RETURN
116 PRINT "FREQ     F(MAG)       F(THETA)    FREQ     F(MAG)       F(THETA)"
118 GOTO 112
137 REM  ******************************
138 REM  *        PRINT OUTPUT        *
139 REM  ******************************
140 IF COR$ = "P" AND M = 2 THEN GOSUB 170
141 FOR Z = 0 TO Q / 2 - 1
142 PRINT USING "##_     "; Z;
144 PRINT USING "+###.#####_    "; C(M, Z); S(M, Z);
145 PRINT USING "##_     "; (Z + Q / 2);
146 PRINT USING "+###.#####_    "; C(M, Z + Q / 2); S(M, Z + Q / 2)
147 NEXT Z
148 RETURN
```

Positive Frequency Stretching

```
200 REM ******************************
202 REM *     TRANSFORM/RECONSTRUCT    *
204 REM ******************************
206 FOR J = 0 TO Q2: REM SOLVE EQNS FOR POSITIVE FREQUENCIES
208 FOR I = 0 TO Q - 1: REM MULTIPLY AND SUM EACH POINT
210 C(M,J) = C(M,J) + C(N,I)*COS(J*I*K1) + K6*S(N,I)*SIN(J*I*K1)
211 S(M,J) = S(M,J) - K6*C(N,I)*SIN(J*I*K1) + S(N,I)*COS(J*I*K1)
212 NEXT I
214 C(M, J) = C(M, J) / K5: S(M, J) = S(M, J) / K5: REM SCALE RESULTS
216 NEXT J
218 RETURN
220 REM ******************************
222 REM *         PLOT FUNCTIONS       *
224 REM ******************************
225 SFF = 4: SFT = 64
226 SCREEN 9, 1, 1: COLOR 9, 1: CLS : YF = -1: YT = -1
228 LINE (0, 5)-(0, 155): LINE (0, 160)-(0, 310)
230 LINE (0, 155)-(600, 155): LINE (0, 235)-(600, 235)
232 GOSUB 266
234 COLOR 15, 1
236 FOR N = 0 TO Q - 1
238 GOSUB 260
240 LINE (X, Y)-(X, Y): LINE (X, Z)-(X, Z)
242 NEXT N
244 LOCATE 2, 10: PRINT "FREQUENCY DOMAIN (MAG)"
246 LOCATE 20, 15: PRINT "TIME DOMAIN"
248 LOCATE 24, 1
250 INPUT "C/R TO CONTINUE"; A$
252 SCREEN 0, 0, 0
254 RETURN
256 REM ******************************
260 Y = C(2, N): Y = 155 - (YF * Y)
262 X = N * 600 / Q: Z = 235 - (YT * C(1, N))
264 RETURN
265 REM ******************************
266 YF = 150/SFF: YT=150/SFT: LINE(0,5)-(5,5): LINE (0,80)-(5,80)
268 LINE (0, 160)-(5, 160): LINE (0, 235)-(5, 235)
270 LOCATE 1, 2: PRINT SFF: LOCATE 6, 2: PRINT SFF / 2
272 LOCATE 12, 2: PRINT "+"; SFT / 2: LOCATE 23, 2: PRINT "-"; SFT / 2
274 RETURN
299 REM ******************************
300 CLS : REM *     STRETCHING THEOREM     *
301 REM ******************************
302 FOR I = 0 TO Q - 1: C(1, I) = 0: S(1, I) = 0
304 FOR J = 1 TO 2: KC(J, I) = 0: KS(J, I) = 0: NEXT: NEXT
305 COR$ = "P": Q = 16: K1 = P2 / Q: Q2 = Q / 2
306 GOSUB 900
308 REM *** GENERATE "Z1" FUNCTION ***
310 PRINT : PRINT SPC(18); " - Z1 - FUNCTION": PRINT
312 C(1, 0) = 8: C(1, 1) = -8: C(1, 2) = 8: C(1, 3) = -8
314 GOSUB 158: REM PRINT HEADING
316 M = 1: GOSUB 140: REM PRINT INPUT FUNCTION
318 PRINT : INPUT "C/R TO CONTINUE"; A$
320 GOSUB 40: REM TAKE XFORM
322 GOSUB 220: REM PLOT DATA
```

Appendix 7.1

```
324 FOR I = 0 TO Q - 1: C(1, I) = 0: S(1, I) = 0: NEXT
326 Q = 32: K1 = P2 / Q: Q2 = Q / 2
328 C(1, 0) = 8: C(1, 2) = -8: C(1, 4) = 8: C(1, 6) = -8
330 GOSUB 158: REM PRINT HEADING
332 M = 1: GOSUB 140: REM PRINT INPUT FUNCTION
334 PRINT : INPUT "C/R TO CONTINUE"; A$
336 GOSUB 40: REM TAKE XFORM
338 GOSUB 220: REM PLOT DATA
396 RETURN
900 CLS : SCREEN 9, 1, 1: COLOR 15, 1: REM TEST DESCRIPTION
902 FOR DACNT = 1 TO 11
904 READ A$: PRINT A$
906 NEXT
908 INPUT "C/R TO CONTINUE"; A$
910 SCREEN 0, 0, 0: RETURN
920 DATA "                    STRETCHING THEOREM TEST"
922 DATA " "
924 DATA "In this illustration we generate a very simple function "
926 DATA "which has two primary characteristics: it is easy to"
928 DATA "generate and it has a distinctive spectrum - it is easy to"
930 DATA "manipulate and easy to recognize.  First we generate the "
932 DATA "function and analyze it (16 data points and 9 frequency "
934 DATA "components).  Then we intersperse zeros and analyze the "
936 DATA "function a second time (now we have 32 data points and 17 "
938 DATA "frequencies). This illustrates the stretching Theorem."
940 DATA " "
1000 END
```

APPENDIX 7.2

SMALL ARRAY FFTS

In Chapter 7 (Fig. 7.4, *p.* 73) we illustrated just how simple small array FFTs were. There are occasions when we only need a small array FFT (e.g., image compression), and the simplicity of small arrays allow for very simple and fast algorithms.

Starting from the PFFFT algorithms on page 75 we have very simple equations for the 2-point DFTs (lines 110-124). We do not solve these equations, however, but substitute them directly into the routines for 4 point DFTs (lines 126-136), yielding:

```
10 C(0) = (Y(0) + Y(2) + Y(4) + Y(6))/4
12 C(1) = (Y(0) - Y(4))/4: S(1) = Y(2) - Y(6))/4
14 C(2) = (Y(0) - Y(2) + Y(4) - Y(6))/4
```

That's all there is to a 4 point DFT. If you compare this algorithm to Fig. 7.4, you will see that it is a direct "lift off" from the 4 point DFT graphic on the left of that figure. You would need to take care that no errors are picked up from the zero frequency sine component (nor from the Nyquist sine component), but then, you hardly need a defined array for these three harmonics—you could just call them FC0, FC1, FS1 and FC2. As a practical matter, you will find it hard to transform 4 point data arrays more efficiently (or quickly) than this.

We can, of course, extend this approach to an 8 point DFT. There would be two sets of the above equation, of course—one for the even data points and one for the odd:

```
10 T(0) = Y(0)+Y(4): T(1) = Y(2)+Y(6)
11 T(2) = Y(0)-Y(4): T(3)=Y(2)-Y(6)
12 C(2) = (T(0)-T(1))/8
13 T(5) = T(0) + T(1)
and:
14 T(6) = Y(1) + Y(5): T(7) = Y(3) + Y(7)
15 T(8) = Y(1) - Y(5): T(9) = Y(3) - Y(7)
16 T(10) = T(6) + T(7)
17 T(11) = T(8) + T(9): T(12) = T(8) - T(9)
```

Note that the divide by 4 operation at the end of each line is removed, except for C(2) and S(2) which are divided by 8.

In the conventional FFT we would combine these odd and even transforms by the familiar rotate and sum *loop*; but, to obtain maximum efficiency, we will combine these via five individual equations:

```
18 K1 = SQR(.5)/8
20 C(0) = (T(5) + T(10))/8
```

```
22 C(1) = T(2) + K1*T(12)
24 S(1) = T(3) + K1*T(11)
26 C(2) = (T(0) - T(1))/8: S(2) = (T(6) - T(7))/8
28 C(3) = T(2) + K1*T(11)
30 S(3) = K1*T(12)-T(3)
32 C(4) = (T(5)-T(10))/8
```

We pick up the division by 4 (eliminated from the previous routines) by dividing by 8 in this routine. The only non-trivial twiddle factor is $\cos(\pi/4) = \sin(\pi/4) = \sqrt{.5}$. Once again, you will find it difficult to beat the efficiency and speed of this routine. You should recognize we're *not* abandoning the fundamental *rotate and sum* mechanism of the FFT—we're only weeding-out superfluous machine cycles.

This approach can be extended to 16 point arrays but it's apparent that, as the twiddle factors become more complicated, it becomes harder to obtain significant gains. It might be a worthwhile exercise to see how far this approach can *profitably* be extended. In any case, you will find it difficult to beat the efficiency and speed of the routines given here....

APPENDIX 9.2

THE BASIC SCHEME OF A FRACTIONAL FREQUENCY POSITIVE FREQUENCY FAST FOURIER TRANSFORM (FF-PFFFT)

We start with an 8 point data array:

$$\left| \begin{array}{c} \text{DATA0} \\ \text{ARRAY} \end{array} \right| = \left| D_0, D_1, D_2, D_3, D_4, D_5, D_6, D_7 \right| \quad \text{-----------} (A9.2.0)$$

The positive frequency 1/2 Hz transform of which will be:

$$\text{Xform} \left| \text{DATA0} \right| = \left| F_0, F_{1/2}, F_1, F_{3/2}, F_2, F_{5/2}, F_3, F_{7/2}, F_4 \right| \quad \text{-----} (A9.2.1)$$

Once again we subtract out the odd data elements creating two stretched arrays:

$$\left| \text{DATA1'} \right| = \left| D_0, 0, D_2, 0, D_4, 0, D_6, 0 \right| \quad \text{---------} (A9.2.2)$$

$$\left| \text{DATA2'} \right| = \left| 0, D_1, 0, D_3, 0, D_5, 0, D_7 \right| \quad \text{--------} (A9.2.2A)$$

This time it works like this: from our discussion of the stretching theorem for positive frequencies we know the transform of (A9.2.2):

$$\text{Xform} \left| \text{DATA1'} \right| = \left| F_0, F_{1/2}, F_1, F_{3/2}, F_2, F_{3/2}^*, F_1^*, F_{1/2}^*, F_0^* \right| \quad \text{------} (A9.2.3)$$

(Where F_1^*, F_0^*, etc., are complex conjugates of F_1, F_0, etc.)

So we remove the zeros in the array $\left| \text{DATA1'} \right|$ and obtain the array:

$$\left| \text{DATA1} \right| = \left| D_0, D_2, D_4, D_6 \right| \quad \text{------------------} (A9.2.4)$$

The half frequency transform of this unstretched array is:

$$\text{Transform} \left| \text{DATA1} \right| = \left| F_0, F_{1/2}, F_1, F_{3/2}, F_2, \right| \quad \text{----} (A9.2.5)$$

Again, F_0, $F_{1/2}$, F_1, etc., in Equation (A9.2.5) are identical to F_0, $F_{1/2}$, etc., in Equation (A9.2.3). Very well then, we will obtain the transform of the stretched data (consisting of 9 frequency components) by finding the transform of a 4 point array (5 frequencies), create the complex conjugate of $F_{3/2}$, F_1, etc., and add them in again as $F_{5/2}$, F_3, etc. By the Addition Theorem we will add the two spectrums (i.e., for odd and even data points) obtained in this manner just as before, rotating the phase of the odd data transform to account for the shifting phenomenon. That is to say, the frequency components from the "odd data" transform are properly phase shifted and summed into the frequency components of the "even data" transform. As before, the virtual components are created by summing the mirror image of the direct components (i.e., the complex conjugates are summed in reverse order).

Writing a fractional frequency transform would be a good exercise to undertake—it would make much of what we have discussed about the FFT come sharply into focus. Here is a little help:

The FFT starts from the *one point xforms* that each data point represents, shifting and adding through N stages to accomplish the transform of Q data points ($Q = 2^N$). Now, in the conventional FFT, we sum each data point from the lower half of the array with the data point that is *half an array* away (i.e., D_0 is summed with $D_{Q/2}$, etc.). In the FFT and PFFFT we start with *single point* xforms and generate, in the first stage, *two point* xforms; however, in a half-frequency PFFFT, we must produce *three* frequency components in the first stage.... In a shifting and adding algorithm, how is the 1/2 Hz component generated?

It will help to return to the basic model for the FFT algorithm discussed in *Understanding the FFT*, and realize that we are shifting the "odd" data points so that they will interleave with the *stretched* "even" data points. The *shifting theorem* has shown that the shift of each component must be proportional to the *frequency*, and a half Hz frequency will only receive half the shift as a full Hz. It may help to review Fig. 7.4 and think of the 4-point xform as a 2-point, half frequency xform, with the last two data points packed with zeros. It may also help to review Fig. 7.6, packing half of the array with zeros.

You recognize, of course, that this is deliberately presented as a puzzle. It's a worthwhile puzzle, even if you do not plan to use the 1/2 Hz configuration. It's apparent that many readers will have neither the time nor inclination to tackle this puzzle...but, quite frankly, it is hard to imagine that, if you have read this far, you will not have the ability. At any rate, if you ever need it, it will be here.

APPENDIX 9.3

ONE DECI-BEL

Just as the *Ohm* is named in honor of Georg Simon Ohm, the *Bel* is named in honor of Alexander Graham Bell.[1] It has a definition:

$$\text{Attenuation (in Bels)} = \log_{10} (P_1/P_2) \qquad (A9.3.1)$$

P_1 and P_2 are *power* readings. This is the *only* definition of the Bel.

By taking the logarithm of the power ratio we can compress a wide dynamic range into a small scale (the range of one megawatt to one microwatt is only 12 Bels). Fortunately, our senses measure magnitude logarithmically, allowing us, for example, to see the Sun (almost) and (later) see very faint stars; consequently, measuring things in Bels frequently fits much better than a linear scale.

The average person, when she wants an evaluation of some man, asks her girlfriend; "How would you rate him on a scale of 0 to 10?" (The *pat* answer, of course, is -1.) Engineers, however, don't want to be like "normal" people (*fat chance*); so, they evaluate everything on a scale of 0 to 100. This requires splitting Bels into *deci-bels* (abbreviated db) which are obviously defined:

$$\text{Attenuation (in db)} = 10 \log_{10}(P_1/P_2) \qquad (A9.3.2)$$

This is still the *same definition*—we're just working with 1/10th units.

Now, power is a good way to measure things—power makes resistors get hot, etc. Electrical engineers, however, don't like to think about resistors getting hot, so they work mostly with *volts*. Be that as it may, it would still be nice to use this logarithmic scale in our work with voltage. This is no problem—if we know the voltage and the resistance, we can always calculate the power:

$$\text{Power (in Watts)} = V^2/R \qquad (A9.3.3)$$

If we substitute this into Equation (9.3.2) we will get:

---

[1] There is, no doubt, a connection here with the notorious "language skills" of engineers...

$$\text{Attenuation (in db)} = 10 \log_{10}[(V_1^2/R_1)/(V_2^2/R_2)] \qquad (A9.3.4)$$

Frequently, the two resistors will be the same value (when we measure two voltages at the output of an amplifier, for example) and then:

$$\text{Attenuation (in db)} = 10 \log_{10}(V_1^2/V_2^2) \qquad (A9.3.5)$$

From our vast knowledge of logarithms we know this equals:

$$\text{Attenuation (in db)} = 20 \log_{10}(V_1/V_2) \qquad (A9.3.6)$$

This is still the same thing as defined in (A9.3.1)!!!

Similarly, the ratio of two currents ($P = I^2R$) is:

$$\text{Attenuation (in db)} = 20 \log_{10}(I_1/I_2) \qquad (A9.3.7)$$

Unfortunately, engineers abuse this formula, and frequently state voltage and current ratios in dbs even when the two resistances are different. Technically, they are wrong; but, in certain situations (when driving high impedance loads with a low impedance source, for example) we can sort-of make sense of this.[2] However... it makes no sense whatsoever when people do dumb things like stating *impedance ratios* in terms of decibels. Now, surely you can plot impedance ratios on a logarithmic scale, or even convert to the logarithms of impedance ratios, but *these are not decibels!... (and multiplying by 20 turns a misdemeanor into a felony)*. We simply *must* draw the line somewhere... otherwise, the definition becomes meaningless.

---

[2] Under these conditions we can imagine some hypothetical load impedance (significantly greater than the internal impedance of the source), across which these voltage readings do indeed indicate legitimate power ratios, and satisfy Alex.

APPENDIX 9.4

SPECTRUM ANALYZER II

We have incorporated a lot of things to simulate a spectrum analyzer, so let's get right to work:

```
' ******************************************************************
' ***   FFT09-02 *** POSITIVE FREQUENCY FFT ***
'THIS PROGRAM ANALYZES TIME DOMAIN DATA WITH FRACTIONAL FREQUENCY
'ANALYSIS. IT SIMULATES THE PERFORMANCE REQUIRED FOR A FREQUENCY
'ANALYZER AND INCLUDES THE LATEST PFFFT (FFT08-01).
' ******************************************************************
10 SCREEN 9, 1: COLOR 15, 1: CLS ' SETUP DISPLAY SCREEN
12 QX = 2 ^ 13: QI = 2 ^ 6: WSF = 1' MAX & NOM SIZE & S.F. CORR
14 N = 12: X0 = 50: Y0 = 10: ASF = 120: SCALE = 1
16 Q = 2^N: N1 = N - 1: Q1 = Q - 1: Q2 = Q/2: Q3 = Q2 - 1: Q4 = Q/4
18 Q5 = Q4 - 1: Q8 = Q / 8: Q9 = Q8 - 1: Q34 = Q2 + Q4: Q16 = Q / 16
20 DIM Y(QX), C(QX), S(QX), KC(Q2), KS(Q2)
22 PI = 3.14159265358979#: P2 = PI * 2: K1 = P2 / Q
24 IOFLG = 2: WTFLG = 1 ' SET TO GRAPHIC DISPLAY AND NO WEIGHTING
26 KLOG = LOG(10): YSF = LOG(ASF)/KLOG: SK1 = 1
28 WEXP = 6: FRACF = 1
32 FOR I = 0 TO Q3: KC(I) = COS(I * K1): KS(I) = SIN(I * K1): NEXT I
34 GOSUB 900 ' SETUP SYSTEM

    '  **************************************************
    '  ********   MAIN MENU (ANALYZER SETUP)   ********
    '  **************************************************
40 CLS : LOCATE 2, 30: PRINT "ANALYZER SETUP MENU"
42 LOCATE 6, 1' DISPLAY MENU
60 PRINT SPC(5); "1 = ANALYZE 64 POINT ARRAY": PRINT
62 PRINT SPC(5); "2 = ANALYZE 128 POINT ARRAY": PRINT
64 PRINT SPC(5); "3 = ANALYZE 256 POINT ARRAY": PRINT
66 PRINT SPC(5); "4 = ANALYZE 512 POINT ARRAY": PRINT
68 PRINT SPC(5); "5 = ANALYZE 1024 POINT ARRAY": PRINT
70 PRINT SPC(5); "6 = ANALYZE 2048 POINT ARRAY": PRINT
72 PRINT SPC(5); "7 = ANALYZE 4096 POINT ARRAY": PRINT
73 PRINT SPC(5); "8 = CHANGE SYSTEM SETUP": PRINT
74 PRINT SPC(5); "9 = END": PRINT
78 PRINT SPC(10); "MAKE SELECTION: ";
80 A$ = INKEY$: IF A$ = "" THEN 80
82 IF ASC(A$)<49 OR ASC(A$)>57 THEN PRINT A$; "= BAD KEY": GOTO 42
90 A = VAL(A$): ON A GOSUB 850,860,865,870,875,880,885,900,999
92 GOTO 40
94 RETURN

       '       *****************************
100    '       ***   FORWARD TRANSFORM   ***
       '       *****************************
```

```
110 C(0) = (S(0) + S(Q2)) / 2: C(1) = (S(0) - S(Q2)) / 2
112 FOR I = 1 TO Q3: I2 = 2*I: INDX = 0 ' BIT REVERSE DATA ADDRESSES
114 FOR J = 0 TO N1: IF I AND 2^J THEN INDX = INDX + 2^(N-2-J)
116 NEXT J
118 C(I2) = (S(INDX)+S(INDX+Q2))/2: C(I2+1) = (S(INDX)-S(INDX+Q2))/2
120 NEXT I
122 FOR I = 0 TO Q1: S(I) = 0: NEXT I

    '        *********  REMAINING STAGES  **********
124 FOR M = 1 TO N1: QP = 2 ^ M: QPI = 2 ^ (N1 - M)
126   FOR K = 0 TO QPI - 1
128     FOR J = 0 TO QP/2: J0 = J+(2*K*QP): J1 = J0+QP: K2 = QPI*J
130       JI = J1 - (2 * J)
132       CTEMP1 = C(J0) + C(J1) * KC(K2) - S(J1) * KS(K2)
134       STEMP1 = S(J0) + C(J1) * KS(K2) + S(J1) * KC(K2)
136       CTEMP2 = C(J0) - C(J1) * KC(K2) + S(J1) * KS(K2)
138       S(JI) = (C(J1) * KS(K2) + S(J1) * KC(K2) - S(J0)) / 2
140       C(J0) = CTEMP1 / 2: S(J0) = STEMP1 / 2: C(JI) = CTEMP2 / 2
142     NEXT J
144   NEXT K
146 NEXT M
148 FOR J = Q2 + 1 TO Q1: C(J) = 0: S(J) = 0: NEXT J
150 T9 = TIMER - T9
152 ON IOFLG GOSUB 300, 350 ' DISPLAY SPECTRUM
154 RETURN

300 '        *******  PRINT OUTPUT  *******
160 FOR Z = 0 TO Q5' PRINT OUTPUT
162 PRINT USING "####"; Z; : PRINT "   ";
164 PRINT USING "+##.#####"; SK1 * C(Z); : PRINT "   ";
166 PRINT USING "+##.#####"; SK1 * S(Z); : PRINT "     ";
168 PRINT USING "####"; Z + Q4; : PRINT "   ";
170 PRINT USING "+##.#####"; SK1 * C(Z + Q4); : PRINT "   ";
172 PRINT USING "+##.#####"; SK1 * S(Z + Q4)
174 NEXT Z
176 PRINT "T = "; T9: INPUT "ENTER TO CONTINUE"; A$
178 RETURN

    ' **********************************
    ' *         PLOT SPECTRUM          *
    ' **********************************
350 CLS : LINE (X0 - 1, 11)-(X0 - 1, Y0 + 320)' DRAW Y AXIS
352 LINE (X0, Y0 + 1)-(X0 + 500, Y0 + 1)' DRAW X AXIS
    ' **** DRAW 20 DB LINES ****
354 FOR I = 2 TO 14 STEP 2: YSKT = INT(YSF*10*LOG(1/(10^I))/KLOG)
355 LINE (X0, Y0 - YSKT)-(X0 + 500, Y0 - YSKT)
356 YDB = CINT(.3 + (Y0 - YSKT) / 13.9): IF YDB > 25 THEN 360
358 LOCATE YDB, 2: PRINT USING "###"; 10 * I;
360 NEXT I
362 YP = SCALE * SQR(C(I) ^ 2 + S(I) ^ 2)    ' FIND RSS OF DATA POINT
```

Spectrum Analyzer II

```
364 IF YP = 0 THEN YP = -160: GOTO 370' OUT OF RANGE, SKIP
366 YP = 20 * LOG(YP)/ KLOG' FIND DB VALUE
370 LINE (X0, Y0 - (YSF*YP))-(X0, Y0 - (YSF*YP))' SET PEN TO ORIGIN
372 FOR I = 0 TO Q3 ' *******   PLOT DATA POINTS    *******
374 YP = SCALE * SQR(C(I) ^ 2 + S(I) ^ 2)   ' FIND RSS OF DATA POINT
380 IF YP = 0 THEN YP = -160: GOTO 384' OUT OF RANGE, SKIP
382 YP = 20 * LOG(YP) / KLOG' FIND DB VALUE
384 LINE -(X0 + XSF * I, Y0 - YSF * YP)' DRAW LINE
386 NEXT I
388 LOCATE 1, 70: PRINT "F = "; : PRINT USING "###.#"; F8
390 RETURN

    ' ***********************************
    ' *      GENERATE SINE WAVE        *
    ' ***********************************
400 FOR I = 0 TO QDT: C(I) = 0: S(I) = SIN(F9 * K1 * I): NEXT
402 FOR I = QDT+1 TO Q: C(I) = 0: S(I) = 0: NEXT
404 IF FLG80 = 1 THEN GOSUB 410
406 IF WTFLG = 2 THEN 450
408 RETURN
410 FOR I = 0 TO QDT: S(I) = S(I) + .0001*SIN(2*F9*K1*I): NEXT
412 RETURN
450 ' ****  WEIGHTING FUNCTION  ***
452 FOR I = 0 TO QDT
454 S(I) = S(I) * (SIN(I * PI / QDT) ^ WEXP)
456 NEXT I
458 RETURN

    ' ***********************************
600 ' ***     SPECTRUM ANALYZER      ***
    ' ***********************************
602 CLS : PRINT : PRINT
604 PRINT SPC(20); "PREPARING DATA - PLEASE WAIT"
610 GOSUB 400 ' GENERATE SINUSOID
620 GOSUB 100 ' ANALYZE SPECTRUM
624 REPT = 0  ' RESET REPEAT FLAG
626 LOCATE 24, 65: PRINT "RETURN TO EXIT";
628 A$ = INKEY$: IF A$ = "" THEN 628' WAIT USER INPUT
630 IF ASC(A$) = 0 THEN GOSUB 650' CURSOR HAS LEADING ZERO
632 IF REPT = 1 THEN 620 ' ANALYZE SPECTRUM AGAIN
634 RETURN ' BACK TO MAIN MENU
650 ' ***  HANDLE CURSOR KEYS  ***
652 A = ASC(RIGHT$(A$, 1))  ' WHICH CURSOR
654 IF A < 75 OR A > 77 OR A = 76 THEN 669 ' NOT A CURSOR KEY
656 IF A = 75 THEN F8 = F8 - .1 ' INC FREQUENCY
658 IF A = 77 THEN F8 = F8 + .1 ' DEC. FREQUENCY
660 F9 = F8 * Q / QI' SCALE FOR CURRENT ARRAY SIZE
662 GOSUB 400 ' GENERATE NEW SINUSOID
664 REPT = 1 ' SET REPEAT FLAG
669 RETURN ' DO IT AGAIN SAM
```

Appendix 9.4

```
       ' ************************************************
800    ' *      SETUP FRACTIONAL FREQUENCY ANALYZER      *
       ' ************************************************
850  N = 6: N1 = 5: Q = 2 ^ N  ' SET ARRAY SIZE
852  QI = Q / FRACF: Q2 = Q / 2: Q3 = Q2 - 1: Q4 = Q / 4: Q5 = Q4 - 1
853  Q8 = Q / 8: Q9 = Q8 - 1: Q16 = Q / 16: Q34 = Q2 + Q4
854  F8 = 16: F9 = F8 * Q / QI: K1 = P2 / Q
855  QDT = Q / FRACF - 1  ' NEW TWIDDLES NEXT LINE
856  FOR I = 0 TO Q3: KC(I) = COS(K1 * I): KS(I) = SIN(K1 * I): NEXT
857  XSF = 500/Q2: SCALE = WSF * FRACF * 2
858  GOSUB 600  ' ANALYZE SPECTRUM
859  RETURN  ' BACK TO MAIN MENU
860  N = 7: N1 = 6: Q = 2 ^ N
862  GOTO 852
865  N = 8: N1 = 7: Q = 2 ^ N
867  GOTO 852
870  N = 9: N1 = 8: Q = 2 ^ N
872  GOTO 852
875  N = 10: N1 = 9: Q = 2 ^ N
877  GOTO 852
880  N = 11: N1 = 10: Q = 2 ^ N
882  GOTO 852
885  N = 12: N1 = 11: Q = 2 ^ N
887  GOTO 852

       ' ************************
       ' *     SYSTEM SETUP     *
       ' ************************
900  CLS : RTFLG = 1: PRINT SPC(20); "      SYSTEM SETUP MENU"
902  PRINT : LOCATE (5): PRINT "1 = DISPLAY "
904  PRINT : PRINT "2 = WEIGHTING FUNCTION"
906  PRINT : PRINT "3 = FRACTIONAL FREQUENCY"
907  PRINT : PRINT "4 = -80 DB COMPONENT"
908  PRINT : PRINT "5 = EXIT"
910  A$ = INKEY$: IF A$ = "" THEN 910
912  A = ASC(A$): IF A < 49 OR A > 53 THEN 900
914  A = A - 48: ON A GOSUB 920, 930, 970, 988, 990
916  ON RTFLG GOTO 900, 928
       ' ************************
920  CLS
922  PRINT "USE GRAPHIC DISPLAY (Y/N)";
924  A$ = INKEY$: IF A$ = "" THEN 924
926  IF A$ = "N" OR A$ = "n" THEN IOFLG = 1 ELSE IOFLG = 2
928  RETURN
       ' ************************
930  CLS : PRINT "WEIGHTING FUNCTION ON (Y/N)";
932  A$ = INKEY$: IF A$ = "" THEN 932
934  IF A$ = "N" OR A$ = "n" THEN WTFLG = 1: WSF = 1: GOTO 956
936  WTFLG = 2: PRINT
938  PRINT "CHANGE WEIGHTING FUNCTION EXPONENT?"
```

Spectrum Analyzer II

```
940 A$ = INKEY$: IF A$ = "" THEN 940
942 IF A$ = "N" OR A$ = "n" THEN 952
944 PRINT "1 = SIN^2": PRINT "2 = SIN^4": PRINT "3 = SIN^6"
946 A$ = INKEY$: IF A$ = "" THEN 946
948 A = ASC(A$): IF A < 49 OR A > 51 THEN 946
950 A = A - 48: WEXP = 2 * A
952 WSF = 2: IF A = 2 THEN WSF = 8 / 3
954 IF A = 3 THEN WSF = 16 / 5
956 RETURN

 ' ************************
970 CLS : PRINT : PRINT "SELECT FRACTIONAL FREQUENCY FOR ANALYSIS"
972 PRINT : PRINT "1 = 1/1"; SPC(20); "4 = 1/8"
974 PRINT "2 = 1/2"; SPC(20); "5 = 1/16"
976 PRINT "3 = 1/4"; SPC(20); "6 = 1/32"
978 A$ = INKEY$: IF A$ = "" THEN 978
980 A = ASC(A$): IF A < 49 OR A > 54 THEN 978
982 A = A - 49: FRACF = 2 ^ A
984 RETURN

 ' ************************
988 FLG80 = 1 - FLG80: RETURN
990 RTFLG = 2: RETURN

 ' ************************
999 STOP ' THAT'S ALL FOLKS
```

Except for a few constants necessary to make this program run, the first part of the program is the same as FFT9.01 The Main Menu (array size selection) is now located at line 40, and now the fraction to be packed with zeros is variable (we will need this to experiment). Selection 8 gets us to the System Set-up, where we find another menu.

1. Weighting Function - Selection 2 jumps to line 930. Line 934 checks if we are turning *weighting OFF*. If so, both WTFLG and WSF are set to 1 and we exit. *Note: You will find the WTFLG in the generate sinusoid routine (line 406) where it causes a jump to the weighting sub-routine (line 450) only if it equals 2. WSF is used in the Plot Data routine (line 372) where it corrects the scale factor for the attenuation introduced by the weighting function.* If we want weighting ON we set the WTFLG = 2 (line 936) and then ask if the weighting exponent is to be changed (see p. 103 in text). Three options are offered at line 944, and we calculate WEXP at line 950 and set WSF before exiting this routine. *Note: WEXP is used in the Weighting sub-routine at line 454.*

2. Selection of the fractional frequency configuration is performed

in the sub-routine at line 970. At line 980 the value entered is checked for a valid selection, then the fraction denominator FRACF is calculated (line 982). *Note: FRACF is used in the Array Selection (line 852) to set QI. It's also used in line 855 to set QDT (i.e., the length of the actual data), and also in the Plot Data routine (to set the scale factor—line 857).*

    3. We need to talk about the Generate Sinewave Function routine (line 400). At line 412 we generate *two* sine waves—the second is 80 db down and separated by 50% of the primary functions frequency (you may want to change this). Also, if the weighting function is selected, we jump to the Weighting sub-routine at line 406. Weighting is handled (lines 452 through 456) by multiplying the data by a half cycle sine wave raised to the power WEXP (see *p.* 103).

    4. We must also talk about the Plot Data routine (line 350-390), but we we calculate the X and Y scale factors at line 857 (where they are determined). At line 350-352 we draw Y and X axes. The data is now displayed in a logarithmic format and, when we draw 20 db attenuation marks (354-360), we must calculate their logarithms (line 354). Likewise, when we plot the data (lines 372-386) we must caluclate the logarithm of that (line 382).

    5. Finally, this program is intended for experimentation—to get a feel for the various configurations and what they can do. It's limited in speed and size but it can still be helpful in gaining an understanding of how spectrum analysis is performed with the FFT and the problems to be expected. It's also intended for experimentation with the code...and possible routines.... For example, you might want to experiment with various weighting functions so don't hesitate to make variations.

APPENDIX 10.0

NEGATIVE FREQUENCIES

A10.1 NEGATIVE FREQUENCIES

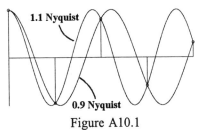

Figure A10.1

Why (you have probably asked yourself) would sinusoids with frequencies equally above and below Nyquist give identical results? Let's look at a modified version of Figure 10.4 which plots sinusoids both 10% above the Nyquist and 10% below. As you might already have guessed, both these sinusoids have the *same value at the sampling points*—but why? Well, a cosine wave is symmetrical about any argument of $N\pi$ (where N is any integer including zero). That is, starting at the peak value (positive or negative) of a cosine, and moving either backward *or* forward (by equal amounts), yields the same value (Fig. A10.2). If the *sample times* occur precisely at $t_S = 2N/f_N$ (i.e., at twice the Nyquist frequency), then a cosine wave 10% above this frequency will advance by precisely the amount that

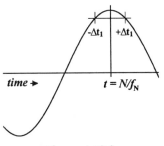

Figure A10.2

a cosine 10% below will be retarded. That is, the argument of a sinusoid 10% above Nyquist will be:

$$2\pi(1.1f_N)t_S = 2\pi(f_N + 0.1f_N)t_S = 2\pi f_N(t_S + 0.1t_S) \quad (A10.1)$$

and the argument of a sinusoid 10% below the Nyquist will be:

$$2\pi(0.9f_N)t_S = 2\pi f_N(t_S - 0.1t_S) \quad (A10.2)$$

Cosines at these arguments must give identical values—sinusoids really are simple things.

On the other hand, *sine* waves (as opposed to *cosine* waves) whose frequencies are equally above and below the Nyquist do *not* yield identical values when sampled at $t_S = 2N/f_N$— they yield values that are the *negative* of each other (Fig. A10.3). So, for the general case of complex sinusoids, if those below the Nyquist are legitimate, their illegitimate equivalents (i.e., those above the Nyquist), if they are to yield the same digitized value, must have negated sine components (but the cosine components remain positive)....

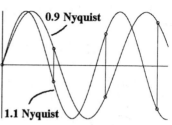

Figure A10.3

Now, any time we deal with complex sinusoids of this form (i.e., identical cosine components but sine components that are the negatives of each other) we're dealing with complex conjugates (you remember these from *p*.10). These come about when the *arguments* of the sinusoids are identical but one is negative, and *that* can happen only when either time or frequency is negative. So, what's negative frequency...or negative time for that matter?

When they launch a rocket over at the *Cape* they always count *down*: "T *minus* 5 minutes and counting"...and later, "T *minus* 10, 9, 8,... 3, 2, 1...*Liftoff!*" Actually, they're not counting down—they're counting *up* in *negative time*. Negative time results when we pick some arbitrary point in time (regardless of whether it's in the future, past, or the present) and call it "T-0." Let's look at what happens:

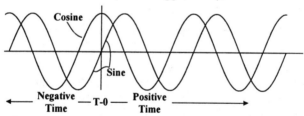

Figure A10.4 - Sinusoids in Positive and Negative Time

If we plot Cos(ωt) and Sin(ωt) about t = 0, they will necessarily look like the sinusoids of Fig. A10.4—time to the left of T-0 is negative and time to the right is positive. Now, when we "count down" to launch our rocket, we have no choice but to count... "T-3, -2, -1, ..."; however, when we trace sinusoids back in negative time, we invariably think of *increasing* numerical values in the negative direction. In any case, this negative argument yields a cosine wave that has the same value at any point in negative time that it has in positive time—and a sine wave that has *negative values* for corresponding points in negative time. This is written:

Negative Frequencies

and:[1]
$$Cos(2\pi t) = Cos(2\pi f(-t)) \quad (A10.3)$$
$$Sin(2\pi ft) = -Sin(2\pi f(-t)) \quad (A10.4)$$

Now, from Fig. A10.5 below, we see that when we deal with a complex sinusoid, the relationship we have just described still holds for the

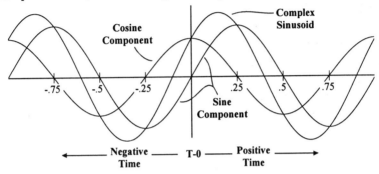

Figure A10.5 - Complex Conjugation

sine and cosine components of that wave; so, clearly a negative time argument yields the complex conjugate of the positive time argument (that is, in terms phasors, the negative time domain waveform will be generated by the complex conjugate of the positive time phasor).

Now, when we deal with the harmonics above the Nyquist in the FFT, it's obvious we're *not* dealing with negative time; however, it's not obvious we're dealing with negative frequency—we certainly haven't *deliberately* introduced negative frequencies. In Equations (A10.1) and (A10.2) above, however, we see that, for each sampling pulse, the argument steps *forward* by half a cycle of the Nyquist signal frequency ± *the difference between Nyquist and the sinusoid being digitized*:

$$2\pi f_N(t_N \pm 0.1 t_N) \quad (A10.5)$$

---

[1] In both of these cases there is a symmetry, although it's obviously not the same kind of symmetry. The symmetry exhibited by the cosine wave is called *even* while the symmetry displayed by the sine wave is called *odd*. If the harmonic components of a function are composed of only sine waves, the whole function will display this *odd* symmetry (i.e., for negative arguments the function itself will be equal to the negative of its positive counterpart). Functions composed of only cosine components will exhibit *even* symmetry. A square wave constructed of sine waves, for example, exhibits odd symmetry.] Functions containing both sine *and* cosine components (i.e., complex sinusoids) will exhibit neither form of symmetry discussed above; nonetheless, the *odd* and *even* condition still holds for the sine and cosine *components* of these functions.

Relative to the Nyquist frequency then, one of these sinusoids is moving positively and the other is moving negatively (see Figs. A10.1 and A10.3). It's as if, from sample to sample, the argument of one is negative (relative to the Nyquist frequency)—the result is identical to that obtained if we were using a negative argument; but, no matter whether we look at this mathematically or otherwise, we're moving forward in positive time! The digitized data is indistinguishable from data generated via a *negative frequency* sinusoid. What we have done here (via the sampling mechanism) is establish the Nyquist sampling rate as the *zero reference frequency*—just as we establish an arbitrary point in time as *T-0*. In sampled data, frequencies below the sample rate must be negative while frequencies above must be positive.

APPENDIX 11.1

OVERSAMPLE DEMONSTRATION

We must make the following changes to the basic PFFFT of Chapter 8. Replace everything up to line 100 with:

```
' *** FFT11.01 - Q (=2^N) POINT FFT (POSITIVE FREQUENCIES ONLY) ***
' ILLUSTRATES OVERSAMPLING - ALSO SEE FFT11.02
10 SCREEN 9, 1: COLOR 15, 1: CLS 'SETUP DISPLAY SCREEN
12 CLS : PRINT "INPUT NUMBER OF DATA POINTS AS 2^N. "
14 INPUT "N = "; N
16 Q = 2^N: N1 = N - 1: Q1 = Q - 1: Q2 = Q/2: Q3 = Q2 - 1: Q4 = Q/4
18 Q5 = Q4 - 1: Q8 = Q/8: Q9 = Q8 - 1: Q34 = Q2 + Q4: Q16 = Q/16
20 DIM C(Q*16), S(Q*16), KC(Q2*16), KS(Q2*16)' DIM DATA
30 PI = 3.141592653589793#: P2 = 2*PI: K1 = P2/Q
    ' *** GENERATE TWIDDLE FACTORS ****
32 FOR I = 0 TO Q2: KC(I) = COS(K1*I): KS(I) = SIN(K1*I): NEXT
34 IOFLG = 2' SET OUTPUT TO GRAPHIC DISPLAY
    ' ********  MAIN MENU  ********
40 CLS ' DISPLAY MAIN MENU
50 PRINT SPC(30); "MAIN MENU": PRINT : PRINT
60 PRINT SPC(5); "1 = ANALYZE COSINE FUNCTION": PRINT
62 PRINT SPC(5); "2 = INVERSE TRANSFORM": PRINT
64 PRINT SPC(5); "3 = SETUP DISPLAY": PRINT
65 PRINT SPC(5); "4 = OVERSAMPLE DATA": PRINT
66 PRINT SPC(5); "5 = EXIT": PRINT
70 PRINT SPC(10); "MAKE SELECTION :";
80 A$ = INKEY$: IF A$ = "" THEN 80
90 A = VAL(A$): ON A GOSUB 600, 200, 800, 700, 990
95 GOTO 40
```

Change line 610 to:

```
610 FOR I = 0 to Q1: C(I) = 0: S(I) = COS(F9 * F1 * I): NEXT I
```

Then, at line 700, we add:

```
700 REM *** 16X OVERSAMPLE DATA ***
702 FOR I = Q2 TO Q ' CLEAR UPPER ARRAY
704 C(I) = 0: S(I) = 0
706 NEXT I
710 N = N + 4: N1 = N - 1' INCREASE 2^N EXPONENT
714 Q = 2 ^ N: Q1 = Q - 1' INCREASE Q & GEN NEW CONSTANTS
716 Q2 = Q / 2: Q3 = Q2 - 1: Q4 = Q / 4: Q5 = Q4 - 1: Q8 = Q / 8
718 K1 = P2 / Q
    ' *** GENERATE NEW TWIDDLE FACTORS ****
720 FOR I = 0 TO Q2: KC(I) = COS(K1 * I): KS(I) = SIN(K1 * I): NEXT
722 GOSUB 350 ' PLOT OVERSAMPLED DATA
```

```
724 INPUT A$ ' WAIT USER - DATA IS NOW OVERSAMPLED
726 RETURN

800 REM ***   SETUP PRINTOUT/GRAPHIC DISPLAY   ***
802 CLS ' CLEAR SCREEN
804 INPUT "GRAPHIC DISPLAY (Y/N)"; A$
806 IF A$ = "Y" THEN IOFLG = 2 ELSE IOFLG = 1
808 RETURN

990 END: STOP
```

This illustration generates a sinusoid and takes the transform automatically. It might be a good idea to reconstruct this signal just to verify the expected *intermodulation* waveform is obtained. Next, generate the sinusoid again; but, this time, execute the oversampling routine before reconstructing. Let's look at the specifics of the program.

Note that, at line 20, we dimension the arrays for 16 times their normal size. This is necessary to handle the oversampled spectrum size. At line 702-706 we clear the upper portion of the data arrays. At line 710 we simply increase the value of N by 4 and at 714 we generate the new value of Q (thereby increasing the array size by 16x). At 716 we re-define all the constants used in the PFFFT that refer to the array size and, at line 720, we generate new twiddle factors. We display this newly oversampled spectrum and that is all there is to it. When we perform the inverse transform we will be reconstructing a time domain array that has 16 times as many data points in the same original interval.

You might want to change the oversampling multiplier to 8x, 4x, etc., to get a feel for what these multipliers do.

The *problem* with this technique may be illustrated by simply running the above demonstration and selecting a frequency of $F = 4.5$. We will talk about *handling* this difficulty in the next appendix.

APPENDIX 11.2

OVERSAMPLING II

We may attempt to solve the problem of oversampling fractional frequencies by using a weighting function. The following routines must be added to the program of the preceding appendix:

```
' *** FFT11.02 - Q (=2^N) POINT FFT (POSITIVE FREQUENCIES ONLY) ***
' ILLUSTRATE OVERSAMPLING ALSO SEE FFT11.02
10 SCREEN 9, 1: COLOR 15, 1: CLS 'SETUP DISPLAY SCREEN
12 CLS : PI = 3.141592653589793#: P2 = 2 * PI
14 IOFLG = 2' SET OUTPUT TO GRAPHIC DISPLAY
16 PRINT "INPUT NUMBER OF DATA POINTS AS 2^N. "
18 GOSUB 86
40 CLS ' DISPLAY MAIN MENU
50 PRINT SPC(30); "MAIN MENU": PRINT : PRINT
60 PRINT SPC(5); "1 = ANALYZE SINE FUNCTION": PRINT
62 PRINT SPC(5); "2 = INVERSE TRANSFORM": PRINT
64 PRINT SPC(5); "3 = SETUP DISPLAY": PRINT
65 PRINT SPC(5); "4 = OVERSAMPLE DATA": PRINT
66 PRINT SPC(5); "5 = EXIT": PRINT
70 PRINT SPC(10); "MAKE SELECTION :";
80 A$ = INKEY$: IF A$ = "" THEN 80
82 A = VAL(A$): ON A GOSUB 600, 200, 800, 700, 990
84 GOTO 40
' **************
86 INPUT "N = "; NI
88 N = NI: Q = 2 ^ N: K1 = P2 / Q: N1 = N - 1: Q1 = Q - 1
90 Q2 = Q / 2: Q3 = Q2 - 1: Q4 = Q / 4: Q5 = Q4 - 1: Q8 = Q / 8
92 REDIM C(8 * Q), S(8 * Q), KC(4 * Q), KS(4 * Q) ' DIM DAT & TWIDDLE
93 ' ***  GENERATE TWIDDLE FACTORS  ****
94 FOR I = 0 TO Q2: KC(I) = COS(K1 * I): KS(I) = SIN(K1 * I): NEXT
96 RETURN

400 ' GENERATE SINUSOID
410 FOR I = 0 TO Q1
416 C(I) = 0: S(I) = COS(F9 * K1 * I)
418 NEXT
432 GOSUB 450 ' INVOKE WEIGHTING FUNCTION
434 GOSUB 370 ' LOOK AT WEIGHTED FUNCTION
436 INPUT A$ ' AWAIT USER'S PLEASURE
440 RETURN

450 ' WEIGHTING FUNCTION
452 FOR I = 0 TO Q1
454 S(I) = S(I) * (SIN(I * PI / Q) ^ 6)
456 NEXT I
458 RETURN
```

```
470 ' INVERSE WEIGHTING FUNCTION
472 FOR I = 1 TO Q1
474 S(I) = S(I) / (SIN(I * PI / Q) ^ 6)
476 NEXT I
478 RETURN

600 CLS : PRINT : PRINT ' GENERATE COSINE COMPONENT
601 GOSUB 88 ' RESET ARRAY SIZE
602 INPUT "PLEASE SPECIFY FREQUENCY "; F9
604 PRINT "PREPARING DATA INPUT - PLEASE WAIT!"
610 GOSUB 400 ' GENERATE COSINE WAVE
620 GOSUB 100 ' PERFORM PFFFT
630 RETURN

700 REM *** 8X OVERSAMPLE DATA ***
702 FOR I = Q2 TO Q ' CLEAR UPPER ARRAY
704 C(I) = 0: S(I) = 0
706 NEXT I
710 N = N + 3: N1 = N - 1' INCREASE 2^N EXPONENT
712 Q = 2 ^ N: Q1 = Q - 1' INCREASE Q & GEN NEW CONSTANTS
716 Q2 = Q / 2: Q3 = Q2 - 1: Q4 = Q / 4: Q5 = Q4 - 1: Q8 = Q / 8
718 K1 = P2 / Q
719 ' *** GENERATE NEW TWIDDLE FACTORS ****
720 FOR I = 0 TO Q2: KC(I) = COS(K1 * I): KS(I) = SIN(K1 * I): NEXT
722 GOSUB 350 ' DISPLAY OVERSAMPLED DATA
724 INPUT A$ ' WAIT USER - DATA IS NOW OVERSAMPLED
726 RETURN
```

In addition to these changes we must add a line at the end of the Inverse Transform routine:

```
243 GOSUB 470 ' PERFORM INVERSE WEIGHTING
```

This program will now automatically invoke the weighting function when the data is generated...and remove it when the data is reconstructed. You will notice that we have placed the initialization of the data array constants in a separate sub-routine (lines 86-96). This eliminates the need to restart the program after each demonstration (see line 601).

Note that, in the Inverse Weighting routine, we do not divide the first and last data points by the weighting function. This would obviously involve dividing by zero.

APPENDIX 11.3

DIGITAL AUDIO ANALYZER

This program is a modification of the spectrum analyzer we developed in Appendix 9.4. Unfortunately it has been modified to the point that the original configuration is obscured.

```
' ****************************************************************
'          *** FFT11-03 *** POSITIVE FREQUENCY FFT ***
'THIS PROGRAM SIMULATES A SPECTRUM ANALYZER USED TO ANALYZE DIGITAL
'AUDIO FOR NON-LINEAR DISTORTION. IT IS BASED ON THE ANALYZER OF
'FFT09-02 AND INCLUDES THE LATEST PFFFT (FFT08-01).
' ****************************************************************
10 SCREEN 9, 1: COLOR 15, 1: CLS ' SETUP DISPLAY SCREEN
12 QX = 2 ^ 13: QI = 2 ^ 6: WSF = 1' MAX & NOM SIZE & S.F. CORR
14 N = 12: X0 = 50: Y0 = 10: ASF = 224
16 Q = 2 ^ N: N1 = N - 1: Q1 = Q - 1: Q2 = Q/2: Q3 = Q2 - 1: Q4 = Q/4
18 Q5 = Q4 - 1: Q8 = Q / 8: Q9 = Q8 - 1: Q34 = Q2 + Q4: Q16 = Q/16
20 DIM Y(QX), C(QX), S(QX), KC(Q2), KS(Q2)
22 PI = 3.14159265358979#: P2 = PI * 2: K1 = P2 / Q: NYQ = 22050
24 IOFLG = 2: WTFLG = 1 ' SET TO GRAPHIC DISPLAY AND NO WEIGHTING
26 XSF = 500 / Q2: KLOG = LOG(10): YSF = LOG(ASF) / KLOG: SK1 = 1
28 WEXP = 6: FRACF = 1 ' WEIGHTING EXP = 6 & FRAC FREQ = 1/1
30 AMP = 1: MAMP = 2 ^ 15: OVSAP = 2: T12 = 1.0594631#
32 GOSUB 900 ' SETUP SYSTEM
34 GOSUB 885 ' SET UP INITIAL ARRAY SIZE OF 2^12
36 GOTO 600 ' START ANALYZER
98 '          *****************************
100 '         ***   FORWARD TRANSFORM    ***
102 '         *****************************
106 '         ***   TRANSFORM STAGE 1    ***
110 C(0) = (S(0) + S(Q2)) / 2: C(1) = (S(0) - S(Q2)) / 2
112 FOR I = 1 TO Q3: I2 = 2*I: INDX = 0 ' BIT REVERSE DATA ADDRESSES
114 FOR J = 0 TO N1: IF I AND 2^J THEN INDX = INDX + 2^(N - 2 - J)
116 NEXT J
118 C(I2) = (S(INDX)+S(INDX+Q2))/2: C(I2+1) = (S(INDX)-S(INDX+Q2))/2
120 NEXT I
122 FOR I = 0 TO Q1: S(I) = 0: NEXT I
'         ********* REMAINING STAGES **********
124 FOR M = 1 TO N1: QP = 2 ^ M: QPI = 2 ^ (N1 - M)
126   FOR K = 0 TO QPI - 1
128     FOR J = 0 TO QP/2: J0 = J + (2*K*QP): J1 = J0 + QP: K2 = QPI*J
130       JI = J1 - (2 * J)
132       CTEMP1 = C(J0) + C(J1) * KC(K2) - S(J1) * KS(K2)
134       STEMP1 = S(J0) + C(J1) * KS(K2) + S(J1) * KC(K2)
136       CTEMP2 = C(J0) - C(J1) * KC(K2) + S(J1) * KS(K2)
138       S(JI) = (C(J1) * KS(K2) + S(J1) * KC(K2) - S(J0)) / 2
140       C(J0) = CTEMP1 / 2: S(J0) = STEMP1 / 2: C(JI) = CTEMP2 / 2
142     NEXT J
144   NEXT K
146 NEXT M
148 FOR J = Q2 + 1 TO Q1: C(J) = 0: S(J) = 0: NEXT J
152 GOSUB 350 ' DISPLAY SPECTRUM
154 RETURN
```

```
    ' ************************************
    ' *         PLOT SPECTRUM            *
    ' ************************************
350 CLS : LINE (X0 - 1, Y0)-(X0 - 1, Y0 + 330)' DRAW Y AXIS
352 LINE (X0, Y0 + 1)-(X0 + 500, Y0 + 1)' DRAW X AXIS
    '       **** DRAW 20 DB LINES ****
354 FOR I = 2 TO 14 STEP 2: YSKT = INT(YSF*10*LOG(1/(10^I))/KLOG)
356 LINE (X0, Y0 - YSKT)-(X0 + 500, Y0 - YSKT)
358 YDB = CINT(.4 + (Y0 - YSKT) / 15.666): IF YDB > 25 THEN 362
360 LOCATE YDB, 2: PRINT USING "###."; 10 * I;
362 NEXT I
364 YP = SCALE*SQR(C(0)^2 + S(0)^2) ' FIND RSS OF DATA POINT
366 IF YP = 0 THEN YP = -330: GOTO 370' OUT OF RANGE, SKIP
368 YP = 20 * LOG(YP) / KLOG' FIND DB VALUE
370 LINE (X0, Y0 - YSF * YP)-(X0, Y0 - YSF * YP)' SET PEN TO ORIGIN
372 FOR I = 0 TO Q3 ' *******   PLOT DATA POINTS   *******
374 YP = SCALE * SQR(C(I) ^ 2 + S(I) ^ 2)    ' FIND RSS OF DATA POINT
380 IF YP = 0 THEN YP = -330: GOTO 384' OUT OF RANGE, SKIP
382 YP = 20 * LOG(YP) / KLOG' FIND DB VALUE
384 LINE -(X0 + XSF * I, Y0 - YSF * YP)' DRAW LINE
386 NEXT I
388 LOCATE 1, 66: PRINT "F = "; : PRINT USING "#####.###"; F8
390 RETURN
    ' ************************************
    ' *       GENERATE SINE WAVE         *
    ' ************************************
400 FOR I = QDT TO Q: C(I) = 0: S(I) = 0: NEXT
404 FOR I = 0 TO QDT: C(I) = 0
408 Y = INT(MAMP * AMP * SIN(F9 * K1 * I)) / MAMP
410 S(I) = Y
412 NEXT I
414 IF WTFLG = 2 THEN 450 ' USE WEIGHTING FUNCTION
420 RETURN
450 ' ****  WEIGHTING FUNCTION  ***
452 FOR I = 0 TO QDT
454 S(I) = S(I) * (SIN(I * PI / QDT) ^ WEXP)
456 NEXT I
458 RETURN
    ' ************************************
600 ' ***    SPECTRUM ANALYZER    ***
    ' ************************************
602 CLS : PRINT : PRINT
604 INPUT "PLEASE SPECIFY STARTING FREQUENCY"; F8
606 F9 = F8 * FRACF * Q2 / NYQ
610 GOSUB 400 ' GENERATE SINUSOID
612 GOSUB 100 ' ANALYZE SPECTRUM
614 REPT = 0 ' RESET REPEAT FLAG
620 LOCATE 1, 40: PRINT "ESC TO CHANGE SYSTEM:";
622 A$ = INKEY$: IF A$ = "" THEN 632 ' EXIT OR RUN?
624 IF ASC(A$) = 0 THEN GOSUB 650' CURSOR HAS LEADING ZERO
626 IF ASC(A$) = 27 THEN GOSUB 900
632 IF REPT = 1 THEN 612 ' ANALYZE SPECTRUM AGAIN
634 GOTO 606 ' BACK TO MAIN MENU
650 ' ***  HANDLE CURSOR KEYS  ***
652 A = ASC(RIGHT$(A$, 1)) ' WHICH CURSOR
```

Digital Audio Analyzer

```
654 IF A < 72 OR A > 80 THEN 664 ' NOT A CURSOR KEY
656 IF A = 72 THEN AMP = AMP * 10: IF AMP > 1 THEN AMP = 1
658 IF A = 75 THEN F8 = F8 - .0625 ' INC FREQUENCY
660 IF A = 77 THEN F8 = F8 + .0625 ' DEC. FREQUENCY
662 IF A = 80 THEN AMP = AMP / 10: IF AMP < .00001 THEN AMP = .00001
663 GOSUB 670
664 F9 = F8 * FRACF * Q2 / NYQ' SCALE FOR CURRENT SYSTEM
666 GOSUB 400 ' GENERATE NEW SINUSOID
668 REPT = 1: RETURN ' SET REPEAT FLAG AND REPEAT
670 IF F8 < NYQ / (OVSAP * FRACF) THEN 676
672 LOCATE 10,10: INPUT "FREQ OUT OF RANGE - ENTER TO CONTINUE"; B$
674 INPUT "PLEASE SELECT STARTING FREQ"; F8
676 RETURN

    ' ***************************************************
    ' ********   ARRAY SIZE MENU (ANALYZER SETUP)   ********
    ' ***************************************************
800 CLS : LOCATE 2, 30: PRINT "ANALYZER SETUP MENU"
802 LOCATE 6, 1' DISPLAY MENU
810 PRINT SPC(5); "1 = ANALYZE 64 POINT ARRAY": PRINT
812 PRINT SPC(5); "2 = ANALYZE 128 POINT ARRAY": PRINT
814 PRINT SPC(5); "3 = ANALYZE 256 POINT ARRAY": PRINT
816 PRINT SPC(5); "4 = ANALYZE 512 POINT ARRAY": PRINT
818 PRINT SPC(5); "5 = ANALYZE 1024 POINT ARRAY": PRINT
820 PRINT SPC(5); "6 = ANALYZE 2048 POINT ARRAY": PRINT
822 PRINT SPC(5); "7 = ANALYZE 4096 POINT ARRAY": PRINT
826 PRINT SPC(5); "9 = EXIT MENU": PRINT
828 PRINT SPC(10); "MAKE SELECTION: ";
830 A$ = INKEY$: IF A$ = "" THEN 830
832 IF ASC(A$)<49 OR ASC(A$)>57 THEN PRINT "INVALID KEY": GOTO 830
840 A = VAL(A$):
842 ON A GOSUB 850, 860, 865, 870, 875, 880, 885, 900, 990
844 RETURN

    ' *    SETUP FRACTIONAL FREQUENCY ANALYZER    *
850 N = 6: N1 = 5: Q = 2 ^ N ' SET ARRAY SIZE
852 QI = Q / FRACF: Q2 = Q / 2: Q3 = Q2 - 1: Q4 = Q / 4: Q5 = Q4 - 1
853 Q8 = Q / 8: Q9 = Q8 - 1: Q16 = Q / 16: Q34 = Q2 + Q4
854 F8 = 16: F9 = F8 * Q / QI: K1 = P2 / Q
585 QDT = Q / FRACF - 1 ' NEW TWIDDLES NEXT LINE
856 FOR I = 0 TO Q3: KC(I) = COS(K1 * I): KS(I) = SIN(K1 * I): NEXT
857 XSF = 500 * OVSAP / (QI * FRACF): SCALE = WSF * FRACF * 2
858 RETURN ' BACK TO MAIN MENU

860 N = 7: N1 = 6: Q = 2 ^ N: GOTO 852
865 N = 8: N1 = 7: Q = 2 ^ N: GOTO 852
870 N = 9: N1 = 8: Q = 2 ^ N: GOTO 852
875 N = 10: N1 = 9: Q = 2 ^ N: GOTO 852
880 N = 11: N1 = 10: Q = 2 ^ N: GOTO 852
885 N = 12: N1 = 11: Q = 2 ^ N: GOTO 852
    ' ***********************
    ' *    SYSTEM SETUP    *
    ' ***********************
900 CLS : RTFLG = 1: PRINT SPC(20); "     SYSTEM SETUP MENU"
902 PRINT : LOCATE (5): PRINT "1 = SUMMARY    5 = ARRAY SIZE"
```

```
904 PRINT : PRINT "2 = WEIGHTING FUNCTION    6 = SET FREQUENCY"
906 PRINT : PRINT "3 = FRACTIONAL FREQUENCY"
907 PRINT : PRINT "4 = EXIT MENU             9 = TERMINATE PROGRAM"
908 PRINT : PRINT
910 A$ = INKEY$: IF A$ = "" THEN 910
912 A = ASC(A$): IF A < 49 OR A > 57 THEN 900
914 A = A - 48: ON A GOSUB 920,930,970,990,800,960,928,928,999
916 XSF = 500 * OVSAP/(QI * FRACF): SCALE = 2* WSF*FRACF
918 ON RTFLG GOTO 900, 928
    ' ***   SHOW SYSTEM CONFIGURATION   ***
920 CLS
922 PRINT SPC(20); "SYSTEM SUMMARY": PRINT : PRINT
923 PRINT "FRACTIONAL FREQUENCY = 1/"; FRACF
924 PRINT "WEIGHTING FUNCTION = SIN^"; WEXP; " - WEIGHTING IS ";
925 IF WTFLG = 1 THEN PRINT "OFF" ELSE PRINT "ON"
926 PRINT "ARRAY SIZE IS "; Q: PRINT
927 INPUT "ENTER TO CONTINUE "; A$
928 RETURN

930 CLS : PRINT "WEIGHTING FUNCTION ON (Y/N)?";
932 A$ = INKEY$: IF A$ = "" THEN 932
934 IF A$ = "N" OR A$ = "n" THEN WTFLG = 1: WSF = 1: GOTO 956
936 WTFLG = 2: PRINT ' TURN WEIGHTING ON
938 PRINT "CHANGE WEIGHTING FUNCTION EXPONENT?"
940 A$ = INKEY$: IF A$ = "" THEN 940
942 IF A$ = "N" OR A$ = "n" THEN 956 ' EXIT
944 PRINT "1 = SIN^2": PRINT "2 = SIN^4": PRINT "3 = SIN^6"
946 A$ = INKEY$: IF A$ = "" THEN 946
948 A = ASC(A$): IF A < 49 OR A > 51 THEN 946
950 A = A - 48: WEXP = 2 * A
952 WSF = 2: IF A = 2 THEN WSF = 8 / 3
954 IF A = 3 THEN WSF = 16 / 5
956 RETURN

    ' ***   SET SAMPLING FREQUENCY   ***
960 INPUT "ENTER FREQUENCY OF SAMPLING RATE (IN SPS)"; FSAP
962 NYQ = FSAP / 2' NYQUIST = HALF SAMPLING RATE
964 RETURN

970 CLS : PRINT : PRINT "SELECT FRACTIONAL FREQUENCY FOR ANALYSIS"
972 PRINT : PRINT "1 = 1/1"; SPC(20); "4 = 1/8"
974 PRINT "2 = 1/2"; SPC(20); "5 = 1/16"
976 PRINT "3 = 1/4"; SPC(20); "6 = 1/32"
978 A$ = INKEY$: IF A$ = "" THEN 978
980 A = ASC(A$): IF A < 49 OR A > 54 THEN 978
982 A = A - 49: FRACF = 2 ^ A
984 A = N-5: ON A GOSUB 850,860,865,870,875,880,885
986 RETURN

990 RTFLG = 2: RETURN
    ' **********
999 END: STOP ' THAT'S ALL FOLKS
```

The standard PFFFT runs from line 100 - 150—the inverse transform is not used. We set up a more or less arbitrary configuration

Digital Audio Analyzer

through line 30 and then GOSUB to line 900 to allow setup modification as desired. At line 34 we jump to the array size setup and then jump to 600 to start the analyzer. At line 604 we get the initial frequency, F9 gets the frequency in terms of harmonic number, and we are ready to jump to the Generate Function routine at line 400. This Generate Function routine is discussed in the text (*p.* 141).

Lines 400 - 458 generate the simulated D/A signal. The *function weighting* is invoked as a user option. Returning to line 612 we then jump down to the PFFFT, and the return-jump (to the Plot Transformed Data routine—lines 350-390) takes place from the PFFFT. The Plot Data routine has changed slightly—all of the scale factors etc., are calculated when they are determined so that they need not be calculated each time we plot the data. Aside from these changes it's pretty much the display routine used in Appendix 9.4.

Returning to line 614 we set the Repeat flag to zero and locate the cursor to the upper right-hand corner of the screen. At line 620 we print the note that the Escape key may be used to modify the system configuration. Let's look at how the user controls the spectrum analyzer.

Line 622 checks to see if a key has been pressed—if not the program continues on. If a key *has* been pressed, we check to see if it's a *control key* (line 624). If so we jump to the Handle Control Key routine at line 650, but we will discuss this shortly. Line 626 checks to see if an Escape key has been pressed (ASCII code 27). If so we jump to line 900 where a System Menu is displayed. The number 1 selection will display the current status of the system (i.e., the fractional frequency and weighting function configuration, as well as the size of the data array being used). This subroutine is located at lines 920-928. *[Note that these sub-routines all return to the System Menu.]*

Option 2 allows changing the weighting function parameters (at lines 930-956). The weighting function may be turned "On" or "Off" and the exponent of the 1/2 cycle sine^N weighting function may be selected.

Option 3 allows changing the fractional frequency configuration (lines 970-986). This routine jumps to the select the array *size* routine again (this is necessary to re-establish the array constants).

Option 4 is the Exit back to the Analyzer.

Option 5 allows selecting the array size which is the same routine used in the previous spectrum analyzers.

Option 6 allows changing the sampling rate (lines 960-964).

Option 9 is the orderly way to terminate this program.

If, back at line 622, a cursor key was entered, the program flow jumps to line 650. We note that the control keys (i.e., the cursors) allow changing the amplitude of the signal being analyzed via the *Up* and *Down* cursors (i.e., ↑ and ↓ on the keyboard) which is an interesting illustration (see lines 630-646). The *left* and *right* (i.e., → and ←) keys increment/decrement the frequency. You may recognize T12 (line 30) as the step ratio for the chromatic scale.

    The reader should be warned this program is a conglomeration of several test routines and, while it works, it will probably contain bugs. Handling of scale factors, etc., is sloppy. It would be a good exercise to write a similar program concentrating on a simple, consistent set of amplitude, time and frequency scale factors.

APPENDIX 12.1

PLAYBACK SPEEDUP

Again, we do not show the PFFFT (lines 100-152) and the Inverse Transform (lines 200-244) to save space.

```
/ ****************************************
10 ' *** (FFT12.01) VOICE SPEEDUP FFT ***
/ ****************************************
' DYNAMIC
12 CLS : SK1 = 1: SK3 = 1: Q = 2 ^ 11: Q2 = Q / 2
14 PATHI$ = "C:\": PATHO$ = PATHI$
16 DIM C(Q), S(Q), KC(Q2), KS(Q2), HDR(44)
18 FOR I = 1 TO 44: READ HDR(I): NEXT I
20 DATA 82,73,70,70,187,23,0,0,87,65,86,69,102,109,116,32,16,0,0,0,1
22 DATA 0,1,0,17,43,0,0,17,43,0,0,1,0,8,0,100,97,116,97,151,23,0,0
24 PI = 3.141592653589793#: P2 = 2 * PI: K1 = P2 / Q
30 Q3 = Q2 - 1
32 FOR I = 0 TO Q3: KC(I) = COS(I * K1): KS(I) = SIN(I * K1): NEXT I
34 INPUT "INPUT DATA SAMPLE RATE"; DSR1
36 CLS : GOSUB 1100 ' GET DATA FILE
40 CLS
50 PRINT SPC(30); "MAIN MENU": PRINT : PRINT
52 PRINT SPC(5); "1 = FFT": PRINT
54 PRINT SPC(5); "2 = INV FFT": PRINT
56 PRINT SPC(5); "3 = DISPLAY MENU": PRINT
58 PRINT SPC(5); "4 = SPEEDUP ROUTINE": PRINT
60 PRINT SPC(5); "5 = FILE UTILITY": PRINT
62 PRINT SPC(5); "6 = FILTER": PRINT
68 PRINT SPC(5); "9 = EXIT": PRINT
70 PRINT SPC(10); "MAKE SELECTION :";
80 A$ = INKEY$: IF A$ = "" THEN 80
82 IF ASC(A$) < 49 OR ASC(A$) > 57 THEN 40
90 A = VAL(A$): ON A GOSUB 600,700,800,400,1000,3000,990,990,990
95 IF A = 9 THEN 999
97 GOTO 40
/ ******************************
/ *      VOICE SPEEDUP        *
/ ******************************
400 CLS : PRINT : PRINT
402 PRINT SPC(20); "VOICE SPEEDUP MENU": PRINT : PRINT
404 PRINT SPC(5); "1 = STRETCH SIDEBANDS      2 = FRACTIONAL STRETCH"
405 PRINT SPC(5); "9 = EXIT"
406 A$ = INKEY$: IF A$ = "" THEN 406
407 ON VAL(A$) GOSUB 450, 412, 410, 410, 410, 410, 410, 410, 410
410 RETURN
412 ' *** FRACTIONAL STRETCH ***
414 CLS : INPUT "SELECT CENTER FREQUENCY"; FCO1
415 FCO1 = FCO1 * Q / DSR1
416 PRINT SPC(20); "SELECT SPEEDUP FRACTION": PRINT : PRINT
417 PRINT SPC(5); "1 = -3/4       5 = +1/4"
418 PRINT SPC(5); "2 = -1/2       6 = +1/2"
419 PRINT SPC(5); "3 = -1/4       7 = +3/4"
420 PRINT SPC(5); "4 = 0          8 = N/A"
421 PRINT SPC(5); "9 = EXIT"
```

```
422 A$ = INKEY$: IF A$ = "" THEN 422
423 IF ASC(A$) < 49 OR ASC(A$) > 57 THEN 414
424 A = VAL(A$): IF A = 9 GOTO 449
425 A = A - 4: IF A > 3 THEN A = 3
426 FCO1 = CINT(FCO1): DELF = 4 - A
' SELECT SIDEBANDS
427 I2 = -1: FCO4 = CINT(FCO1 / 8): IX = INT(FCO1 * DELF / 8)
428 FOR I = 0 TO IX STEP DELF: I2 = I2 + 1
429 C(FCO4 + I2) = C(FCO1 + I): S(FCO4 + I2) = S(FCO1 + I)
430 C(FCO4 - I2) = C(FCO1 - I): S(FCO4 - I2) = S(FCO1 - I)
431 NEXT I
432 FOR I = FCO4 + I2 TO Q - 1: C(I) = 0: S(I) = 0: NEXT I
433 N = N - 3: N1 = N - 1: Q = 2 ^ N: K1 = P2 / Q
434 Q2 = Q / 2: Q3 = Q2 - 1: Q4 = Q / 4: Q5 = Q4 - 1: Q8 = Q / 8
436 FOR I = 0 TO Q2: KC(I) = COS(K1 * I): KS(I) = SIN(K1 * I): NEXT I
449 RETURN
450 ' ******   STRETCH SIDEBANDS   *****
452 INPUT "CARRIER FREQUENCY IS"; FCO1
454 FCO1 = FCO1 * Q / DSR1
456 FOR I = 100 TO 0 STEP -1: I2 = 2 * I: I3 = I2 - 1
460 C(FCO1 + I2) = C(FCO1 + I): S(FCO1 + I2) = S(FCO1 + I)
462 C(FCO1 + I3) = 0: S(FCO1 + I3) = 0
464 C(FCO1 - I2) = C(FCO1 - I): S(FCO1 - I2) = S(FCO1 - I)
466 C(FCO1 - I3) = 0: S(FCO1 - I3) = 0
468 NEXT I
470 RETURN
600 ' ***************************************
'   *            FORWARD FFT            *
'   ***************************************
602 CLS : SK1 = 1: K6 = 1: KFB = 2
604 LOCATE 12, 20: PRINT "PLEASE WAIT - TRANSFORMING DATA"
620 GOSUB 100' DO PFFFT
622 CLS : GOSUB 2500' PLOT DATA
624 RETURN
'       **********************************
700 '***           INVERSE FFT          ***
'       **********************************
710 K6 = -1: KFB = 1
712 CLS : LOCATE 12, 20: PRINT "PLEASE WAIT - XFORMING DATA."
714 GOSUB 200' DO INVERSE PFFFT
716 CLS : GOSUB 2000' PLOT DATA
718 RETURN
'           **************************
800 '   *** DISPLAY DATA MENU ***
'           **************************
802 CLS : PRINT SPC(30); "DISPLAY MENU ": PRINT : PRINT
804 PRINT SPC(5); "1 = DISPLAY TIME DOMAIN (GRAPHIC)"
806 PRINT SPC(5); "2 = DISPLAY FREQ DOMAIN (GRAPHIC)"
808 PRINT SPC(5); "9 = EXIT"
810 A$ = INKEY$: IF A$ = "" THEN 810
812 IF ASC(A$) < 49 OR ASC(A$) > 57 THEN 802
814 ON VAL(A$) GOSUB 2000, 2500, 816, 816, 816, 816, 816, 816, 816
816 RETURN
990 CLOSE: RETURN
999 END
```

Playback Speedup

```
'       **************************
'       *        FILE MENU       *
'       **************************
1000 CLS : PRINT SPC(30); "FILE HANDLER MENU ": PRINT : PRINT
1002 PRINT SPC(5); "1 = RELOAD FILE        2 = SAVE FILE"
1004 PRINT SPC(5); "3 = LOAD NEW FILE      4 = CHANGE SAMPLE RATE"
1006 PRINT SPC(5); "5 = WEIGHTING FUNCTION"
1010 PRINT SPC(5); "9 = EXIT"
1012 A$ = INKEY$: IF A$ = "" THEN 1012
1014 ON VAL(A$) GOSUB 1110,1200,1100,1040,1060,1018,1018,1018,1018
1018 RETURN
'       ********  CHANGE INPUT PATH  ************
1020 NPATH$ = ""
1022 INPUT "NEW PATH"; NPATH$
1024 IF NPATH$ = "" THEN 1024
1026 PATHI$ = NPATH$
1028 RETURN
'********* CHANGE OUTPUT PATH   ************
1030 NPATH$ = ""
1032 INPUT "NEW PATH"; NPATH$
1034 IF NPATH$ = "" THEN 1034
1036 PATHO$ = NPATH$
1038 RETURN
'       ************  CHANGE SAMPLE RATE  ************
1040 NRATE = 0
1042 INPUT "NEW SAMPLE RATE"; NRATE
1044 IF NRATE = 0 OR NRATE > 44100 THEN 1042
1046 DSR1 = NRATE: TOO = Q / DSR1
1048 RETURN
'       ********  WEIGHTING FUNCTION  **********
1050 FOR I = 0 TO Q1
1052 S(I) = S(I) * (SIN(I * PI / Q) ^ 6)
1054 NEXT I
1056 RETURN
'       *********  WEIGHTING ON/OFF  ************
1060 PRINT "WEIGHTING ON (Y/N)?";
1062 A$ = INKEY$: IF A$ = "" THEN 1062
1064 IF A$ = "Y" OR A$ = "y" THEN WTFLG = 1
1066 IF A$ = "N" OR A$ = "n" THEN WTFLG = 0
1068 IF A$ <> "Y" AND A$ <> "y" AND A$ <> "N" AND A$ <> "n" THEN 1060
1070 RETURN
'       ********************
'       ***  LOAD FILE  ***
'       ********************
1100 CLS : NUFLG = 1
1102 PRINT "FILENAME ('PATH' TO CHANGE PATH)"; PATHI$;
1104 INPUT NMI$
1106 IF NMI$ = "PATH" THEN GOSUB 1020: GOTO 1100
1108 NMI$ = PATHI$ + NMI$
1109 IF RIGHT$(NMI$, 4) <> ".WAV" THEN NMI$ = NMI$ + ".WAV"
1110 OPEN "R", #1, NMI$, 1
1112 FIELD #1, 1 AS DA$
1114 QDF = LOF(1): QF = QDF - 44
1116 GET #1, 23                 'GET NUMBER OF CHANNELS
1118 CHNLS = ASC(DA$): IF CHNLS = 2 THEN QF = QF/2  '(MONO?STEREO?
```

```
1120 GET #1, 35
1122 WRDSIZ = ASC(DA$): IF WRDSIZ = 16 THEN QF = QF / 2' TWO BYTES
1126 PRINT QF; " AFTER WORDSIZE & # CHANNELS"
1128 GET #1, 26 ' GET SAMPLE RATE HIGH BYTE
1129 SR1 = ASC(DA$)
1130 SR1 = 256 * SR1 ' FILE S/R
1132 GET #1, 25
1134 SR1 = SR1 + ASC(DA$): DSR1 = SR1
1136 CLOSE 1
1138 STZ = 1
1142 IF NUFLG = 0 THEN 1152
1144 NUFLG = 0
1145 PRINT "FREE MEMORY ="; FRE(-1)
1146 PRINT "8 * QF = "; 8 * QF; " PROCEED?";
1148 A$ = INKEY$: IF A$ = "" THEN 1148
1150 IF A$ = "N" OR A$ = "n" THEN 1194
1152 FOR N = 10 TO 13' FIND SIZE OF REQUIRED ARRAY
1154 IF QF <= 2 ^ N THEN 1158
1156 NEXT N
1157 REPROF = 0: 'QF = 2 ^ (N)
1158 IF A$ = "Q" OR A$ = "q" THEN N = 14
1160 Q = 2^N: N1 = N - 1: Q1 = Q - 1: Q2 = Q/2: Q3 = Q2 - 1: Q4 = Q/4
1161 Q5 = Q4 - 1: Q8 = Q/8: Q9 = Q8 - 1: Q34 = Q2 + Q4: Q16 = Q/16
1162 REDIM C(Q), S(Q)
1164 TOO = Q / DSR1: K1 = P2 / Q
1166 CLOSE 1
     TYPE Dat3
         Amp3 AS INTEGER
     END TYPE
1176 DIM WRD AS Dat3
1180 IF WRDSIZ = 16 THEN GOSUB 1300 ELSE GOSUB 1350
1182 CLOSE #1
1184 FOR I = 0 TO Q
1186 C(I) = 0
1188 NEXT I
1190 IF WTFLG = 1 THEN GOSUB 1050
1194 CLS : GOSUB 2000
1196 RETURN
' *******************************************
' *        SAVE TIME DOMAIN FILE            *
' *******************************************
1200 PRINT " SET GAIN (Y/N)?"
1202 A$ = INKEY$: IF A$ = "" THEN 1202
1204 IF A$ = "Y" OR A$ = "y" THEN GOSUB 1270
1210 HDR(7) = INT((2 * Q + 36) / 65536)
1212 HDR(6) = INT(((2 * Q + 36) - HDR(7) * 65536) / 256)
1214 HDR(5) = INT((2 * Q + 36) - (HDR(7) * 65536) - (HDR(6) * 256))
1216 HDR(43) = INT((2 * Q) / 65536)
1218 HDR(42) = INT(((2 * Q) - (HDR(43) * 65536)) / 256)
1220 HDR(41) = INT((2 * Q) - (HDR(43) * 65536) - (HDR(42) * 256))
1222 HDR(26) = INT(DSR1 / 256)
1224 HDR(25) = INT(DSR1 - (HDR(26) * 256)): DSR2 = 2 * DSR1
1226 HDR(31) = INT(DSR2 / 65536): DSR2 = DSR2 - HDR(31) * 65536
1228 HDR(30) = INT(DSR2 / 256): DSR2 = DSR2 - HDR(30) * 256
1230 HDR(29) = INT(DSR2)
```

Playback Speedup

```
1232 PRINT : PRINT "FILENAME ('PATH' TO CHANGE PATH) "; PATHO$;
1234 INPUT NM2$
1236 IF NM2$ = "PATH" THEN GOSUB 1030: GOTO 1232
1238 NM$ = PATHO$ + NM2$
1240 OPEN "R", #2, NM$, 2
1242 FIELD #2, 2 AS DA2$
1244 FOR I = 1 TO 43 STEP 2
1246 HDR1 = HDR(I) + HDR(I + 1) * 256
1248 IF HDR1<32768 THEN LSET DA2$ = MKI$(HDR1): GOTO 1252
1250 LSET DA2$ = CHR$(HDR(I))+CHR$(HDR(I+1))
1252 PUT #2, (I + 1) / 2
1254 NEXT I
1256 FOR I = 0 TO Q - 1
1258 LSET DA2$ = MKI$(INT(SK3 * S(I)))
1260 PUT #2, (I + 23)
1262 NEXT I
1264 CLOSE #2
1266 RETURN
1270 ' ************ SET GAIN *************
1272 ITST = 0: NGAIN = 0
1274 FOR I = 0 TO Q - 1
1276 IF ABS(S(I)) > ITST THEN ITST = ABS(S(I))
1278 NEXT I
1280 PRINT : PRINT "THE PEAK VALUE IN THE ARRAY IS "; ITST
1282 PRINT "MAXIUM GAIN = "; (4096 / ITST)
1284 INPUT "NEW GAIN = "; NGAIN
1286 IF NGAIN = 0 THEN 1282
1288 SK3 = NGAIN
1290 PRINT "GAIN = "; SK3
1292 INPUT "C/R TO CONTINUE"; A$
1294 RETURN
1300 '*********   LOAD 16 BIT WORDS   ************
1302 OPEN NMI$ FOR RANDOM AS #1 LEN = LEN(WRD)
1303     STZ = CHNLS ': ISTOP = 22'       'STEPSIZE = 2 FOR STEREO
1304     FOR I = 0 TO QF - 1
1305     GET #1, (STZ * I) + 45, WRD    'LOCATE STZ*I +45--SO WE STEP
1306     S(I) = WRD.Amp3                'OVER 2 WORDS FOR STEREO
1307     IF CHNLS = 1 THEN 1310         ' = MONAURAL
1308     GET #1, (STZ * I) + 46, WRD    '+46 FOR STEREO
1309     IF I > Q THEN 1312 ELSE S(I) = S(I) + WRD.Amp3
1310     NEXT I
1312 RETURN
1350 '*********   LOAD 8 BIT BYTES
1352 OPEN "R", #1, NMI$, 1
1354 FIELD #1, 1 AS DA$
1356     FOR I = 0 TO QF - 1 ' STEP STZ
1358     GET #1, (STZ * I) + 45
1360     S(I) = ASC(DA$) - 127
1362     NEXT I
1364 RETURN
' ************************************
' *          PLOT DATA (TIME)
' ************************************
2000 IF REPOF = 1 THEN RETURN
2001 SCREEN 9, 1, 1: COLOR 7, 1: MAGX = 1: X0 = 10
```

```
2002 CLS : SKX1 = 600 * MAGX / Q: SKY1 = 0
2004 FOR I = 0 TO Q - 1
2006 IF ABS(S(I)) > SKY1 THEN SKY1 = ABS(S(I))
2008 NEXT I
2010 SKY1 = 145 / SKY1: Y0 = 170
2012 Y2 = Y0 + 5: Y3 = Y0 - 5
2014 LINE (X0, Y0)-(X0, Y0): COLOR 7, 1
2016 FOR I = 0 TO Q - 1
2018 X5 = X0 + (I * SKX1): Y5 = Y0 - (S(I) * SKY1)
2020 LINE -(X5, Y5)
2022 NEXT I
2024 COLOR 15, 1
2026 LINE (10, Y0 - 150)-(10, 150 + Y0)
2028 LINE (10, Y0)-(610, Y0)
2030 FOR I = 1 TO 10
2032 X2 = 10 + 60 * I
2034 LINE (X2, Y2)-(X2, Y3)
2036 NEXT I
2038 LOCATE 1, 1: PRINT "XFORM TIME = "; INT(100 * T9) / 100; " SEC."
2040 LOCATE 1, 35: PRINT "FILE = "; NMI$
2042 LOCATE 22, 60: PRINT "Tmax ="; T00 / MAGX;
2044 LOCATE 22, 1: PRINT "M = EXPAND: O = OFFSET";
2048 A$ = INKEY$: IF A$ = "" THEN 2048
2050 IF A$ = "M" OR A$ = "m" THEN 2056
2052 IF A$ = "O" OR A$ = "o" THEN 2070
2054 IF ASC(A$) = 27 THEN RETURN ELSE 2002
' EXPAND DATA DISPLAY
2056 IF A$ = "M" THEN MAGX = 2 * MAGX
2058 IF A$ = "m" THEN MAGX = MAGX / 2: IF MAGX < 1 THEN MAGX = 1
2060 GOTO 2002
' OFFSET DATA
2070 IF A$ = "O" AND X0 > (600 * (1 - MAGX) + 10) THEN X0 = X0 - 60
2072 IF A$ = "o" THEN X0 = X0 + 60: IF X0 > 10 THEN X0 = 10
2074 GOTO 2002
' ****************************************
' *        PLOT DATA (FREQUENCY)         *
' ****************************************
2500 IF REPOF = 1 THEN RETURN
2502 CLS : SCREEN 9, 1, 1: COLOR 7, 1
2504 MAGX = 1: OFS = 0: SKY1 = 0
' FIND MAX VALUE IN DISPLAY DATA & SET EQUAL TO SKY1
2506 FOR I = 0 TO Q2: YTST = SQR(C(I)^2 + S(I)^2)
2508 IF YTST > SKY1 THEN SKY1 = YTST
2510 NEXT I
' SET DISPLAY SCALE FACTOR = 200 PIXLS/SKY1
2512 SKY1 = 200 / SKY1
2514 Y0 = 300: SKY2 = SKY1 / 2
2516 SKX1 = MAGX * 600 / Q2: X0 = 10' SET X SCALE FACTOR
2522 GOSUB 2900 ' DRAW DATA
2530 FC = INT(Q5/MAGX + OFS): Y3 = Y0+10: Y2 = Y0-220' CURSOR STUFF
2532 FCD = FC - OFS: X5 = X0 + FCD * SKX1 ' MORE CURSOR STUFF
2534 LINE (X5, Y3)-(X5, Y2) ' DRAW CURSOR
2536 GOSUB 2840 ' DISPLAY PERTINENT DIGITAL DATA
2540 A$ = INKEY$: IF A$ = "" THEN 2540' WAIT FOR USER INPUT
2542 IF A$ = CHR$(27) THEN 2552' ESC = EXIT
```

Playback Speedup

```
2544 IF ASC(A$) = 0 THEN 2590 ' HANDLE CURSOR KEYS
2546 IF A$ = "M" OR A$ = "m" THEN 2700 ' EXPAND FREQUENCY SCALE?
2548 IF A$ = "O" OR A$ = "o" THEN 2750 ' OFFSET FREQUENCY BAND
2550 GOTO 2540 ' SOMEBODY HIT THE WRONG KEY
2552 MAGX = 1: OFSI = 0: RETURN
2590 A = INT((ASC(RIGHT$(A$, 1)) - 70) / 2)' HANDLE CURSORS
2591 IF A < 1 OR A > 5 THEN 2540
2592 GOSUB 2800 ' ERASE PRESENT CURSOR
2594 ON A GOTO 2630, 2600, 2620, 2620, 2640
2596 GOTO 2540
' MOVE CURSOR DOWN ONE
2600 FC = FC - 1: IF FC < 3 THEN FC = 3
2610 FCD = FC - OFS
2612 GOSUB 2840 ' PRINT DATA
2614 LINE (X0 + SKX1 * FCD, Y3)-(X0 + SKX1 * FCD, Y2)' DRAW CURSOR
2618 GOTO 2540
' MOVE CURSOR UP ONE
2620 FC = FC + 1: IF FC > Q2 - 3 THEN FC = Q2 - 3
2626 GOTO 2610 ' NO SENSE DOING THIS TWICE
' MOVE CURSOR UP 10
2630 FC = FC + 10: IF FC > Q2 - 3 THEN FC = Q2 - 3' INCREASE CURSOR
FREQUENCY BY 10
2636 GOTO 2610 ' NOR THREE TIMES
' MOVE CURSOR DOWN 10
2640 FC = FC - 10: IF FC < 3 THEN FC = 3
2644 GOTO 2610 ' DITTO
' ***   EXPAND FREQUENCY SCALE FACTOR BY 2X   ***
2700 IF A$ = "M" AND MAGX < 16 THEN MAGX = 2 * MAGX: OFSI = 2 * OFSI
2702 IF A$ = "m" AND MAGX > 1 THEN MAGX = MAGX/2: OFSI = INT(OFSI/2)
2704 GOTO 2756 ' NO SENSE DOING THIS TWICE EITHER
' ***   OFFSET DISPLAYED SPECTRUM
2750 IF A$ = "O" AND OFSI < MAGX THEN OFSI = OFSI + 1
2754 IF A$ = "o" AND OFSI > 0 THEN OFSI = OFSI - 1
2756 OFS = OFSI * Q2 / MAGX: CLS
2758 GOTO 2514 ' OKAY, REPLOT DATA
' ERASE CURSOR
2800 COLOR 1, 1 ' EVERYTHING BLUE
2802 FOR I = -1 TO 1 ' WHERE IS THAT THING
2804 X5 = X0 + SKX1 * (FCD + I)
2806 LINE (X5, Y3)-(X5, Y2) ' ERASE OLD CURSOR
2808 NEXT I
2810 LINE (X0, Y0)-(X0, Y0)
2812 COLOR 7, 1 ' CHANGE THE COLOR BACK
2814 GOSUB 2900 ' REDRAW DATA
2816 RETURN
' DISPLAY DATA PARAMETERS
2840 COLOR 15, 1
2842 LOCATE 24, 1: PRINT "USE ARROWS TO MOVE CURSOR/ESC TO EXIT";
2844 LOCATE 2, 1: PRINT "XFORM TIME = "; INT(100 * T9) / 100
2846 LOCATE 1, 60: PRINT "Fc ="; FC / T00
2848 LOCATE 2, 60: YDIS = (SKY1 * SQR(C(FC) ^ 2 + S(FC) ^ 2))
2850 PRINT "MAGNITUDE = "; : PRINT USING "###.###"; YDIS / 2
2852 LOCATE 1, 1: PRINT "FILE = "; NMI$
2854 COLOR 7, 1
2856 RETURN
```

```
' DRAW DATA
2900 LINE (X0, Y0)-(X0, Y0)
2901 FOR I = OFS TO (Q2 / MAGX + OFS)
2902 X5 = (I - OFS) * SKX1: Y5 = SKY1 * SQR(C(I) ^ 2 + S(I) ^ 2)
2904 LINE -(X0 + X5, Y0 - Y5)
2906 NEXT I

' DRAW ORDINATE & ABSCISSA
2910 LINE (X0, Y0 - 200)-(X0, Y0)
2912 LINE (X0, Y0)-(600 + X0, Y0)
2920 RETURN

' **************************************
' *          FILTER ROUTINES           *
' **************************************
3000 CLS
3002 PRINT "LOWPASS, BANDPASS OR NOTCH (L/B/N)"
3010 A$ = INKEY$: IF A$ = "" THEN 3010
3012 IF A$ = "N" OR A$ = "n" THEN 3100 ' NOTCH FILTER
3014 IF A$ = "B" OR A$ = "b" THEN 3150 ' BANDPASS
3016 IF A$ <> "L" AND A$ <> "l" THEN 3000 ' IF NOT LOWPASS TRY AGAIN
3018 PRINT "GAUSSIAN OR TRUNCATE (G/T)" ' WHAT KIND OF LOWPASS?
3020 A$ = INKEY$: IF A$ = "" THEN 3020 ' GET USER INPUT
3022 IF A$ = "T" OR A$ = "t" THEN 3200 ' TRUNCATE?
3024 PRINT : PRINT "GAUSSIAN LOW PASS FILTER ROUTINE"
3026 INPUT "INPUT CUTOFF FREQUENCY"; FCO3
3028 FCO3 = FCO3 * TOO: FCO32 = FCO3 * FCO3
3030 FOR I = 0 TO Q / 2 ' FILTER THE DATA
3032 KAT = EXP(-.3453878 * I * I / FCO32) ' EIGEN VALUE COEFFICIENT
3034 C(I) = KAT * C(I): S(I) = KAT * S(I)
3036 C(Q - I) = KAT * C(Q - I): S(Q - I) = KAT * S(Q - I)
3038 NEXT I
3040 RETURN

3100 PRINT : PRINT "NOTCH FILTER ROUTINE"
3102 INPUT "CENTER FREQUENCY & BANDWIDTH"; FCEN, BW4
3104 FCEN = FCEN * TOO: BW4 = BW4 * TOO
3106 FOR I = FCEN - BW4 TO FCEN + BW4
3108 C(I) = 0: S(I) = 0
3110 NEXT I
3112 RETURN

3150 PRINT : PRINT "BANDPASS FILTER ROUTINE"
3152 INPUT "CENTER FREQUENCY & BANDWIDTH"; FCEN, BW4
3154 FCEN = FCEN * TOO: BW4 = BW4 * TOO
3156 FOR I = 0 TO FCEN - BW4
3158 C(I) = 0: S(I) = 0
3160 C(Q - I) = 0: S(Q - I) = 0
3162 NEXT I
3164 FOR I = FCEN + BW4 TO Q2
3166 C(I) = 0: S(I) = 0
3168 C(Q - I) = 0: S(Q - I) = 0
3170 NEXT I
3172 RETURN
```

Playback Speedup 243

```
3200 ' *** TRUNCATE UPPER SPECTRUM ***
3202 PRINT : PRINT "TRUNCATE UPPER SPECTRUM ROUTINE"
3204 INPUT "CUTOFF FREQUENCY"; FCO4
3206 FCO4 = FCO4 * TOO
3208 FOR I = FCO4 TO Q / 2
3210 C(I) = 0: S(I) = 0
3212 C(Q - I) = 0: S(Q - I) = 0
3214 NEXT I
3216 RETURN
```

We load and save audio files with the routines between lines 1000-2000 (which takes up 3½ pages of code). There are two major formats for audio: the *.VOC and *.WAV (where the * represents the file name), but we use only the *.WAV. The file *header* is critical if these files are to be used in other programs, and lines 18-24 read .WAV header data into array HDR(44) [required when we save files later].

We load files at lines 1100-1200. A file name may be specified or the current file may be re-loaded (line 1001). Lines 1102-1104 input the file name [an option to change the path (lines 1500-1540) is provided—the default is for C:\). The file is opened at line 1110 and the size of the data QF determined by subtracting 44 header bytes from the Length Of File (line 1116). The *.WAV format allows saving the data as either 16 bit words or 8 bit bytes, which is indicated by the 35th byte of the header (line 1120). This changes everything, of course, as we will see. The sample rate is indicated by bytes 25 and 26 of the header (lines 1128-1134). If we are re-loading the current file (i.e., NUFLG = 0) we can skip down to 1152. If the size of the file is too large to fit into the maximum array size, there is no sense proceeding, and we can exit at line 1150. [If the file was created at 44,100 samples /sec. but is too large to fit in the maximum array size, we can cut it in half by specifying a sample rate of 22,050 and then loading only every second sample—you must make a decision about aliasing, of course.] If the file will "fit" we go on to select the next largest 2^N file size 1152-56, or if "Q" is input at line 1148 we simply use the largest file size possible. Lines 1170-76 establish the type of data we expect to read from the file. At 1180 we jump to either the 16 bit word or 8 bit byte input routine and load the data into the data array. Lines 1184-88 insure the aft end of the array is zeros and at line 1190 weighting the data is allowed. At line 1194 we display the loaded data, and then our file is loaded and ready for work.

To save a file (lines 1200-1300) we may want to change the "gain" (1200—subroutine at 1270). Again, we can specify the name and path (lines 1232-38); but first, we generate the *header* (lines 1210-20) *file size* information. At 1222-30 we enter the sample rate and at 1240-54 we save the header data.

The plot data routines have a couple of features that must be pointed out: you can *expand* and *offset* the *time domain* display (2044-74) by changing the variables MAGX (lines 2056-60), XO (lines 2070-74), and re-plotting the data (the routines are located at lines 2700-2920). We may exit the display routine via the Esc key.

The *Plot Frequency* routine allows expanding the display by pressing "M" (2546), but you can also "shrink" the display by pressing "m." Since expanding the display will move the upper frequencies out of view, you can *offset* the display ("O" offsets + and "o" offsets - in lines 2548). The cursor can be moved across the data using the *arrow* keys (2544). *Left/right* arrows move the cursor ±1 harmonic and *up/down* arrows move the cursor ±10 (routines at lines 2590-2644).

Voice speedup is done in lines 400-470, but before we get there, the *data* must occupy no more than 1/8 of the total array (you must control sample rate, aliasing, etc.). As an exercise you may digitize a word (your own voice) and break it into its major components (using the bandpass filter routine). You then save the individual *elements* (e.g., as HAR01.wav, HAR02.wav, etc.), recall them one by one, change the playback rate, save these new components, and then recombine them. To change playback rate you must transform an element and use the cursor keys to *find its center frequency*. You then select *Fractional Stretch* (lines 402-08), and enter the component's center frequency (line 414). Select the *speedup fraction* desired (menu 416-22) and the routine then does the fractional stretching (see Ch. 12, *pp.* 151-155). You must then *Inverse Transform* this data (note that the time domain data is automatically expanded to nominal half array—see lines 432-36) and then save this stretched sound element. In FFT12.03 (next appendix) the *File Manager* includes a routine for combining two *.wav files. With a little effort you could recombine these stretched files and hear yourself say the same thing faster and slower. [Note: there's a *.wav file included on the software disk (THISF3C.WAV) for the sound element we expanded in Chapter 12.]

APPENDIX 12.2

SYNTHESIZER

This is only a modification of the playback speedup program of the previous appendix—we need not repeat all of that code here. The differences are as follows:

```
10 ' *** FFT12-03 (SYNTH-01) MUSIC SYNTHESIZER FFT PROGRAM ***
 ' DYNAMIC
12 CLS : SK1 = 1: SK3 = 1: Q = 2 ^ 10: Q2 = Q / 2
14 PATHI$ = "A:\": PATHO$ = PATHI$
16 DIM C(Q), S(Q), HDR(44), CD(1600), SD(1600)
18 FOR I = 1 TO 44: READ HDR(I): NEXT I
20 DATA 82,73,70,70,187,23,0,0,87,65,86,69,102,109,116,32,16,0,0,0,1
22 DATA 0,1,0,34,86,0,0,17,43,0,0,2,0,16,0,100,97,116,97,0,0,1,0
24 PI = 3.141592653589793#: P2 = 2 * PI: K1 = P2 / Q
30 Q2 = Q / 2: Q3 = Q2 - 1
36 CLS : GOSUB 1100 ' GET DATA FILE
40 CLS
50 PRINT SPC(30); "MAIN MENU": PRINT
52 PRINT SPC(5); "1 = FFT": PRINT
54 PRINT SPC(5); "2 = INV FFT": PRINT
56 PRINT SPC(5); "3 = SETUP SYNTHESIZER": PRINT
58 PRINT SPC(5); "4 = SYNTHESIZER": PRINT
60 PRINT SPC(5); "5 = FILE UTILITY": PRINT
62 PRINT SPC(5); "6 = FILTER": PRINT
64 PRINT SPC(5); "7 = DISPLAY": PRINT
66 PRINT SPC(5); "8 = FINE LINE SPECTRUM": PRINT
68 PRINT SPC(5); "9 = EXIT": PRINT
70 PRINT SPC(10); "MAKE SELECTION :";
80 A$ = INKEY$: IF A$ = "" THEN 80
82 IF ASC(A$) < 49 OR ASC(A$) > 57 THEN 40
90 A = VAL(A$): ON A GOSUB 600,700,4000,4080, 1000,3000,800,7000
95 IF A = 9 THEN 999
97 GOTO 40

 ' **********************************************************
 ' *                FORWARD PFFFT                           *
 ' **********************************************************
100 '           *** TRANSFORM STAGE 1 ***
108 T9 = TIMER
110 C(0) = (S(0) + S(Q2)) / 2: C(1) = (S(0) - S(Q2)) / 2
112 FOR I = 1 TO Q3: I2 = 2 * I: INDX = 0
114 FOR J = 0 TO N1: IF I AND 2^J THEN INDX = INDX + 2^(N - 2 - J)
116 NEXT J
118 C(I2) = (S(INDX)+S(INDX+Q2))/2:C(I2+1) = (S(INDX)-S(INDX+ Q2))/2
120 NEXT I
122 FOR I = 0 TO Q1: S(I) = 0: NEXT I
 '      ********* REMAINING STAGES **********
124 FOR M = 1 TO N1: QP = 2 ^ M: QPI = 2 ^ (N1 - M)
126   FOR K = 0 TO QPI - 1
128     FOR J = 0 TO QP/2: J0 = J+(2*K*QP): J1 = J0+QP: K2 = QPI*J*K1
130     J1 = J1 - (2 * J)
132     CTEMP1 = C(J0) + C(J1) * COS(K2) - S(J1) * SIN(K2)
134     STEMP1 = S(J0) + C(J1) * SIN(K2) + S(J1) * COS(K2)
```

```
136  CTEMP2 = C(J0) - C(J1) * COS(K2) + S(J1) * SIN(K2)
138  S(JI) = (C(J1) * SIN(K2) + S(J1) * COS(K2) - S(J0)) / 2
140  C(J0) = CTEMP1/2: S(J0) = STEMP1/2: C(JI) = CTEMP2/2
142  NEXT J
144  NEXT K
146  NEXT M
148  FOR J = Q2 + 1 TO Q1: C(J) = 0: S(J) = 0: NEXT J
150  T9 = TIMER - T9: SK1 = 2
152  RETURN
     '*********************************************************
     '*                   INVERSE TRANSFORM                    *
     '*********************************************************
200  PRINT : T9 = TIMER: SK1 = 1: QP = 2 ^ N1
202  FOR M = N1 TO 1 STEP -1'  LOOP FOR STAGES OF COMPUTATION
204  QP2 = 2^(M): QP = INT(QP/2): QP4 = 2*QP2: QPI = 2^(N1 - M)
206   FOR I = 0 TO Q - (QP2) STEP QP4
208    FOR J = 0 TO QP: KI = J + I: KT = J * QPI * K1: KJ = QP2 + KI
212     MCT = C(J+I) - C(I+QP2-J): MST = S(J+I) + S(I+QP2-J)
214     CTEMP = MCT * COS(KT) + MST * SIN(KT)
216     STEMP = MST * COS(KT) - MCT * SIN(KT)
218     CTEMP2 = (2 * C(J + I)) - CTEMP * COS(KT) + STEMP * SIN(KT)
220     S(KI) = (2 * S(J + I)) - CTEMP * SIN(KT) - STEMP * COS(KT)
222     C(KJ) = CTEMP: S(KJ) = STEMP: C(KI) = CTEMP2
224    NEXT J
226   NEXT I
228  NEXT M
229  '********  FINAL STAGE  ********
230  FOR I = 0 TO Q3: I2 = 2 * I: INDX = 0
232   FOR J = 0 TO N1: IF I AND 2^J THEN INDX = INDX + 2^(N - 2 - J)
234   NEXT J
236   S(INDX) = C(I2) + C(I2 + 1): S(INDX + Q2) = C(I2) - C(I2 + 1)
238  NEXT I
240  FOR I = 0 TO Q1: C(I) = 0: NEXT I
242  T9 = TIMER - T9
244  RETURN
     '**************************************
600  '***          FORWARD FFT          ***
     '**************************************
602  SK1 = 1: K6 = 1: KFB = 2
604  IF REPOF = 0 THEN CLS : LOCATE 12, 20: PRINT "PLEASE WAIT"
620  GOSUB 100'  DO PFFFT
621  LOCATE 12, 20: PRINT "                    "
622  GOSUB 2500'  PLOT DATA
624  RETURN
     '**************************************
700  '***          INVERSE FFT          ***
     '**************************************
710  K6 = -1: KFB = 1
712  IF REPOF = 0 THEN CLS : LOCATE 12, 20: PRINT "PLEASE WAIT"
714  GOSUB 200'  DO INVERSE PFFFT
715  LOCATE 12, 20: PRINT "                    "
716  GOSUB 2000'  PLOT DATA
718  RETURN
```

Music Synthesis

```
     '      ***************************
800  '      *** DISPLAY DATA MENU ***
     '      ***************************
802 CLS : PRINT SPC(30); "DISPLAY MENU ": PRINT : PRINT
804 PRINT SPC(5); "1 = DISPLAY TIME (GRAPHIC)"
806 PRINT SPC(5); "2 = DISPLAY FREQ (GRAPHIC)"
808 PRINT SPC(5); "9 = EXIT"
810 A$ = INKEY$: IF A$ = "" THEN 810
812 IF ASC(A$) < 49 OR ASC(A$) > 57 THEN 802
814 ON VAL(A$) GOSUB 2000, 2500, 816, 816, 816, 816, 816, 816, 816
816 RETURN
    '  ***    HAPPY ENDINGS    ***
990 CLOSE
992 RETURN
999 END

    '       **************************
    '       *      FILE MENU      *
    '       **************************
1000 CLS : PRINT SPC(30); "FILE HANDLER MENU ": PRINT : PRINT
1002 PRINT SPC(5); "1 = RELOAD FILE          2 = SAVE FILE"
1004 PRINT SPC(5); "3 = CHANGE NAME & LOAD   4 = SUM IN 2ND FILE"
1006 PRINT SPC(5); "5 = CHANGE SAMPLE RATE"
1008 PRINT SPC(5); "7 = WEIGHTING FUNCTION"
1010 PRINT SPC(5); "9 = EXIT"
1012 A$ = INKEY$: IF A$ = "" THEN 1012
1014 ON VAL(A$) GOSUB 1110,1200,1100,1400,1040,1018,1060,1018,1018
1018 RETURN
1320 '*********   LOAD 2ND FILE 16 BIT WORDS   *************
1322 OPEN NMI$ FOR RANDOM AS #1 LEN = LEN(WRD)
1323    STZ = CHNLS ': ISTOP = 22'      'STEPSIZE = 2 FOR STEREO
1324    FOR I = 0 TO QF - 1
1325    GET #1, (STZ * I) + 45, WRD    'LOCATE STZ*I +45--SO WE STEP
1326    S(I) = S(I) + WRD.Amp3         'OVER 2 WORDS FOR STEREO
1327    IF CHNLS = 1 THEN 1330         ' = MONAURAL
1328    GET #1, (STZ * I) + 46, WRD    'SAME EXCEPT +46 FOR STEREO
1329    IF I > Q THEN 1332 ELSE S(I) = S(I) + WRD.Amp3
1330    NEXT I
1332 RETURN
1370 '*********    LOAD 2ND FILE 8 BIT BYTES
1372 OPEN "R", #1, NMI$, 1
1374 FIELD #1, 1 AS DA$
1376    FOR I = 0 TO QF - 1 ' STEP STZ
1378    GET #1, (STZ * I) + 45
1380    S(I) = S(I) + ASC(DA$) - 127
1382    NEXT I
1384 RETURN
'***************************
'*  SUM TWO FILES TOGETHER *
'***************************
' ASSUMES FIRST FILE IS IN S(I)
1400 CLS : NUFLG = 1
1402 PRINT "FILENAME ('PATH' TO CHANGE PATH)"; PATHI$;
1404 INPUT NMI$
1406 IF NMI$ = "PATH" THEN GOSUB 1020: GOTO 1402
```

```
1408 NMI$ = PATHI$ + NMI$
1410 OPEN "R", #1, NMI$, 1
1412 FIELD #1, 1 AS DA$
1416 GET #1, 23                   'GET NUMBER OF CHANNELS
1418 CHNLS = ASC(DA$)             'SET # CHANNELS (I.E. MONO/STEREO)
1420 GET #1, 35
1422 WRDSIZ = ASC(DA$)
1426 PRINT QF; " AFTER WORDSIZE"
1428 GET #1, 26 ' GET SAMPLE RATE HIGH BYTE
1430 SR1 = ASC(DA$)
1432 SR1 = 256 * SR1 ' FILE S/R
1434 GET #1, 25
1436 SR1 = SR1 + ASC(DA$)
1438 CLOSE 1
1440 STZ = 1
1442 IF NUFLG = 0 THEN 1450
1444 NUFLG = 0
1446 TOO = Q / DSR1: K1 = P2 / Q
1450 IF WRDSIZ = 16 THEN GOSUB 1320 ELSE GOSUB 1370
1452 CLOSE #1
1454 CLS : GOSUB 2000
1456 IF WTFLG = 1 THEN GOSUB 1050
1458 RETURN
'       ***********************************
4000 ' ***    SETUP MUSIC SYNTHESIZER    ***
'       ***********************************
4002 CLS: PRINT "THIS ROUTINE WILL SAVE THE HARMONICS SELECTED IN"
4004 PRINT "THE ARRAYS CD(X) AND SD(X) TO BE USED LATER IN THE"
4005 PRINT "SYNTHESIS OF A SONG. THIS ALGORITHM WILL SERVE TO"
4006 PRINT "DEMONSTRATE THE BASIC PRINCIPLE INVOLVED IN A HIGH"
4008 PRINT "QUALITY SYNTHESIZER. INPUT BASIC INSTRUMENT SOUND"
4010 INPUT "ELEMENTS BANDWIDTH FOR ALL ELEMENTS"; BAND
4012 BAND = CINT(BAND * TOO)
4014 FOR HAR = 0 TO 10 ' DO FOR 10 HARMONICS
4016 PRINT "FREQUENCY CFO FOR ELEMENT"; HAR;
4018 INPUT CFO
4020 CFO = CINT(CFO * TOO): CF1 = CFO: OFSET = 2 * HAR
4022 FOR TF = CF1 - BAND TO CF1 + BAND ' TRANSFER TO DATA ARRAY
4024 CD(TF - CF1 + (1 + OFSET) * BAND) = C(TF)
4026 SD(TF - CF1 + (1 + OFSET) * BAND) = S(TF)
4028 NEXT TF
4030 CD(OFSET * BAND) = CF1' SAVE HARMONIC FREQUENCY
4032 NEXT HAR ' REBUILD THE PIANO BEFORE WE EXIT!
'    SHIFT HARMONICS TO DESIRED NOTE
4034 FOR I = 0 TO Q: C(I) = 0: S(I) = 0: NEXT' ZERO ARRAY
4036 FOR HAR = 0 TO 10: OFSET = 2 * (HAR)
4038 CF = CD(OFSET * BAND)
4040 FSHO = CINT(CF * ((1.0594631# ^ NOTE) - 1))
4042 'IF HAR = 0 THEN FSHO = 0' DON'T SHIFT THE BOX
4044 FOR I = CF - BAND + 1 TO CF + BAND - 1
4046 C(I + FSHO) = CD(I - CF + BAND * (1 + OFSET))
4048 S(I + FSHO) = SD(I - CF + BAND * (1 + OFSET))
4050 NEXT I
4052 NEXT HAR
4054 RETURN
```

Music Synthesis

```
'      *******************************************
'      *     SYNTHESIZE A SONG (01/12/03)       *
'      *******************************************
'      THIS ROUTINE ASSUMES THE FILE "PIANO.WAV" HAS BEEN LOADED
'      AND THE HARMONICS SAVED IN THE CD(I) AND SD(I) FILES). IT
'      ALSO ASSUMES THE NOTE IS "A". THIS ROUTINE CALCULATES
'      THE NOTE AND ITS DURATION AND APPENDS IT TO FILE "SONGNM$"
'      UNTIL THE "END OF SONG" CODE (I.E., "Z") IS INPUT.
4080 REPOF = 1: CLS : PRINT : PRINT "INPUT SONG? (Y/N)";
4084 A$ = INKEY$: IF A$ = "" THEN 4084
4086 IF A$ = "Y" OR A$ = "y" THEN 6000
4088 RETURN
'      ****   SYNTHESIZER   ****
6000 SCREEN 9, 1, 1: CLS : FILOC = 23
6002 PRINT "ENTER FILENAME FOR SONG (PATH TO CHANGE PATH)"; PATHO$;
6004 INPUT SONGNM$: IF SONGNM$ = "PATH" THEN GOSUB 1030: GOTO 6002
6006 GOSUB 6200 'OPEN FILE AND WRITE HEADER
6008 CLS : COLOR 15, 1: ISG = 1: FFO = 469.693
6012 LOCATE 1, 1: PRINT SONGNM$; "  NOTE"; ISG
6016 LOCATE 3, 1: INPUT "ENTER NOTE AS A, A#, Bb, ETC."; A$
6020 A = ASC(A$): A1 = ASC(RIGHT$(A$, 1))
6021 IF A = 111 THEN O8 = -1: GOTO 6016 ' SHIFT DOWN AN OCTAVE
6022 IF A > 71 THEN 6100 ' CLOSE FILE
6024 IF A < 65 THEN 6012
6026 LOCATE 4, 1
6028 INPUT "ENTER DURATION AS 1/2 NOTE = 2, 1/4 = 4, ETC.)"; B19
6030 B = Q / B19' : PRINT "1/"; B19;
6034 ' MULT BY 2 FOR FULL STEPS, EXCEPT AT B-C AND E-F
6036 A = (A - 65)*2: IF A > 3 THEN A = A - 1: IF A > 7 THEN A = A - 1
6038 IF A > 7 OR O8 = -1 THEN A = A - 12: O8 = 0
6040 IF A1 = 35 THEN A = A + 1 ' SHARP?
6041 IF A1 = 98 THEN A = A - 1' FLAT?
6042 NOTE = A
6044 GOSUB 4034 ' PRODUCE NOTE
6046 LOCATE 20, 30: PRINT "INVERSE TRANSFORM"
6048 GOSUB 700  ' INVERSE TRANSFORM
6049 IF B > Q THEN BREM = B - Q: B = Q
6050 FOR I = 0 TO B
6052 LSET DA2$ = MKI$(CINT(SK3 * S(I)))
6054 PUT #2, (I + FILOC)
6056 NEXT I
6058 FILOC = FILOC + B + 1
6059 IF BREM > 0 THEN GOSUB 6080
6060 LOCATE 20, 30: PRINT "                    "
6062 ISG = ISG + 1
6064 LOCATE 3, 1: PRINT "                                        "
6066 LOCATE 4, 1: PRINT "                                        "
6068 LOCATE ISG, 68: IF ISG > 22 THEN 6070
6069 GOSUB 6072 ELSE LOCATE 24, 1: PRINT : LOCATE 22, 68: GOSUB 6072
6070 GOTO 6012
6072 PRINT A$; ", 1/"; B19: RETURN
6080 LSET DA2$ = MKI$(1024)
6082 FOR I = 0 TO BREM: PUT #2, (I + FILOC): NEXT
6084 FILOC = FILOC + BREM + 1: BREM = 0
6086 RETURN
```

Appendix 12.2

```
'     ***********************************************
'     ****   UPDATE HEADER DATA & CLOSE FILE   ****
'     ***********************************************
6100 FILC = 2 * (FILOC - 23) + 36
6102 HDR(8) = INT(FILC/16777216): FILC = (FILC-(HDR(8)*16777216))
6104 HDR(7) = INT(FILC/65536): FILC = FILC - (HDR(7)*65536)
6106 HDR(6) = INT(FILC/256): FILC = FILC - (HDR(6)*256)
6108 HDR(5) = INT(FILC): FILC = 2 * (FILOC - 23)
6112 HDR(44) = INT(FILC/16777216): FILC = FILC - (HDR(44)*16777216)
6114 HDR(43) = INT(FILC / 65536): FILC = FILC - (HDR(43) * 65536)
6116 HDR(42) = INT(FILC / 256): FILC = FILC - (HDR(42) * 256)
6118 HDR(41) = INT(FILC)
6120 HDR(26) = INT(DSR1 / 256)
6122 HDR(25) = INT(DSR1 - HDR(26) * 256)
6126 DSR2 = 2 * DSR1
6128 HDR(31) = INT(DSR2 / 65536): DSR2 = DSR2 - HDR(31) * 65536
6130 HDR(30) = INT(DSR2 / 256): DSR2 = DSR2 - HDR(30) * 256
6132 HDR(29) = INT(DSR2)
6140 FOR I = 1 TO 43 STEP 2
6142 LSET DA2$ = CHR$(HDR(I)) + CHR$(HDR(I + 1))
6144 PUT #2, (I + 1) / 2
6146 NEXT I
6148 CLOSE #2
6150 REPOF = 0: RETURN
'     ***********************************
'     **   OPEN SONG FILE & WRITE HEADER   *
'     ***********************************
6200 NM$ = PATHO$ + SONGNM$
6202 OPEN "R", #2, NM$, 2
6204 FIELD #2, 2 AS DA2$
6206 FOR I = 1 TO 43 STEP 2
6208 LSET DA2$ = CHR$(HDR(I)) + CHR$(HDR(I + 1))
6210 PUT #2, (I + 1) / 2
6212 NEXT I
6214 RETURN
'     ***********************************
'     ***    FINE LINE SPECTRUM    ***
'     ***********************************
7000 CLS : PRINT "CNTR FREQ, BNDWDTH, STEP 1/N (E.G. 350,30,8)"
7002 INPUT " "; CF, BAND, FFRAC
7004 CLS : SCREEN 9, 1, 1: COLOR 7, 1
7006 Y0 = 300: SKY2=SKY1/2: CF=CINT(CF*TOO): BAND = CINT(BAND*TOO)
7008 SKX1 = 600 / (2 * BAND): X0 = 10' SET X SCALE FACTOR
7010 GOSUB 5100 ' SET DISPLAY SCALE FACTOR = 200 PIXLS/SKY1
7012 LINE (X0, Y0 - 200)-(X0, Y0)' DRAW ORDINATE & ABSCISSA
7014 LINE (X0, Y0)-(600 + X0, Y0)
7016 LINE (X0 + (BAND * SKX1), Y0 - 200)-(X0 + (BAND * SKX1), Y0)
7018 LINE (X0 + SKX1 * BAND, Y0 + 5)-(X0 + SKX1 * BAND, Y0 - 200)
7020 LINE (X0, Y0)-(X0, Y0)
7022 LOCATE 20, 37: PRINT CF / TOO
7024 LOCATE 20, 1: PRINT (CF - BAND) / TOO
7026 LOCATE 20, 70: PRINT (CF + BAND) / TOO
7028 DELF = 1 / FFRAC: X5 = 0: Y5 = 0: YMAX = 0
7030 COLOR 15, 1
7032 FOR F = CF - BAND TO CF + BAND STEP DELF
```

Music Synthesis

```
7034 CTOT = 0: STOT = 0: KPI = P2 * F / (Q)
7036 FOR I = 0 TO Q1
7038 CTOT = CTOT+(S(I)*COS(KPI*I)): STOT = STOT+(S(I)*SIN(KPI*I))
7040 NEXT I
7050 LOCATE 1, 1: PRINT F;
7051 PRINT "COS COMPONENT ="; CTOT/Q, "SIN COMPONENT ="; STOT/Q
7052 CTOT = CTOT/Q: STOT = STOT/Q
' DRAW DATA
7054 X6 = X5: Y7 = Y6
7056 X5 = (F-CF+BAND)*SKX1: Y5 = SQR(CTOT^2 + STOT^2): Y6 = SKY1*Y5
7058 LINE (X0 + X6, Y0 - Y7)-(X0 + X5, Y0 - Y6)
7060 IF Y5 > YMAX THEN YMAX = Y5: XMAX = F
7061 LOCATE 2,1: PRINT "YMAX ="; YMAX, "@ F ="; XMAX / T00
7062 NEXT F
7064 LOCATE 3, 1
7066 INPUT "ENTER TO RETURN"; A$
7068 SKY1 = 300 / YMAX
7070 RETURN
5100 ' ********** CALIBRATE SCALE FACTOR **************
5122 F = CF
5124 CTOT = 0: STOT = 0: KPI = P2 * F / (Q)
5126 FOR I = 0 TO Q1
5128 CTOT = CTOT + S(I)*COS(KPI * I): STOT = STOT + S(I)*SIN(KPI*I)
5130 NEXT I
5132 LOCATE 1, 1: PRINT F,
5134 PRINT "COS COMPONENT ="; CTOT/Q, "SIN COMPONENT ="; STOT/Q
5136 CTOT = CTOT / Q: STOT = STOT / Q
5138 SKY1 = 200 / SQR(CTOT ^ 2 + STOT ^ 2)
5140 PRINT 1 / (SKY1 / 200)
5190 CLS : RETURN
```

We calculate the twiddle factors on the fly so as to eliminate the two twiddle factor arrays. This is apparent in lines 132-138 and 214-220. We include the *File Handler Menu* here since we have added a routine to sum two files together (line 1004)—although we do *not* repeat all the other file subroutines. The major difference between this routine (lines 1400-1458) and the *load file* routine is we will already have one file loaded in the S(I) array (either via reconstruction of a modified transform or loading it with the standard loader). This routine *sums* a second time domain file with the data already in the S(I) array (lines 1326-29 and 1380).

We come then to our music synthesizer. We need a single piano note (preferably concert A) for our *model*. At a sample rate of 22,050 sps you will need about 0.743 sec. of data (i.e., 2^{14} data points). Load this file and transform it to the frequency domain, where you may then analyze the major harmonic *elements* to find their center (i.e., peak amplitude) frequencies—let's talk about these.

First of all, in most instruments there will be frequencies below the fundamental of the tone being sounded (the "box sounds"). These are clearly visible in the spectrum, and you should experiment with including and eliminating them. The *Synthesizer Setup* routine expects 10 harmonic elements (plus the *box*) so make sure you log at least 10. Each of these harmonic *elements* has a number of sidebands, and you may experiment with including more or less of these.

The setup routine will then step through 11 harmonic elements asking for the center frequency of each element (you have just logged these when you analyzed the piano sound above). The routine then takes the number of sidebands you specified and transfers them to the arrays CD(I) and SD(I). It then asks for the next harmonic element *center frequency*, etc., etc. Once these harmonic elements are loaded we are then ready to synthesize our song.

The synthesizer top level routine starts at line 4080, but the synthesizer starts at 6000. This is a very simple routine that allows typing in a song (as notes and durations) which are converted to the piano sounds we stored in the setup routine above (lines 4000 - 4032). The actual generation of the transform domain for each sound is performed by the routine at lines 4034 - 4054. Each harmonic element is shifted by the number of half steps required to produce the desired note.

The notes are entered (line 6016) as A, B, C, etc., with the symbol # indicating sharps and b indicating flats. A half note is entered as 2, a quarter note as 4, etc., and you may terminate the song by typing any letter greater than G.

There is one final routine that has been included: the *Fine Line Spectrum* routine (lines 7000-7070). This is just a fractional frequency DFT (as opposed to FFT). It's quite slow but intended for investigating the fine line spectrum in the neighborhood of critical data. It can be used for finding a *best estimate* of center frequencies for example. The human ear can distinguish difference in frequency on the order of 1/100 Hz, and if notes and harmonics are not accurate to that resolution we will never produce a good synthesizer.

This is only a "starter" program—if you get hooked on this you will need to write many additional routines. For example, we control duration by simply truncating the available note, but you should really use the *speed up* technique of the previous chapter, but there's much more.... In any case you can get a feel for where this sort of technology is heading.

APPENDIX 13.1

THE 2-D FFT

The core 2-D FFT is discussed in the text. We will cover the associated routines required to display 2-D functions here. The routines to *generate* 2-D functions are discussed in the next appendix.

```
' *******************************************
' *******  2DFFT 13.00 - 2D XFORM  *******
' *******************************************
10 SCREEN 9, 1, 1: COLOR 15, 1: CLS
14 INPUT "SELECT ARRAY SIZE AS 2^N.  N ="; N
16 N1 = N - 1: Q = 2 ^ N: Q1 = Q - 1
18 ' $DYNAMIC
20 DIM C(Q, Q), S(Q, Q), KC(Q), KS(Q), DAC(Q, Q), DAS(Q, Q)
30 Q2 = Q / 2: Q3 = Q2 - 1: Q4 = Q / 4: Q5 = Q4 - 1: Q8 = Q / 8
32 PI = 3.141592653589793#: PI2 = 2 * PI: K1 = PI2 / Q: CLVK = 1
' **** TWIDDLE FACTOR TABLE GENERATION ****
40 FOR I = 0 TO Q: KC(I) = COS(K1 * I): KS(I) = SIN(K1 * I)
42 IF ABS(KC(I)) < .0000005 THEN KC(I) = 0 ' CLEANUP TABLE
44 IF ABS(KS(I)) < .0000005 THEN KS(I) = 0
46 NEXT I
48 FOR I = 1 TO Q1: INDX = 0
50 FOR J = 0 TO N1
52 IF I AND 2 ^ J THEN INDX = INDX + 2 ^ (N1 - J)
54 NEXT J
56 IF INDX > I THEN SWAP KC(I), KC(INDX): SWAP KS(I), KS(INDX)
58 NEXT I
' ***********************
70 CLS : PRINT : PRINT : PRINT "             MAIN MENU": PRINT
74 PRINT " 1 = TRANSFORM FUNCTION": PRINT
76 PRINT " 2 = INVERSE TRANSFORM ": PRINT
84 PRINT " 3 = GENERATE FUNCTIONS         ": PRINT
88 PRINT " 4 = EXIT              ": PRINT : PRINT
90 PRINT "          MAKE SELECTION";
92 A$ = INKEY$: IF A$ = "" THEN 92
94 A = VAL(A$): ON A GOSUB 100, 150, 5000, 112
96 IF A = 4 THEN 9999
98 GOTO 70
' *******************************************
' *            XFORM FUNCTION             *
' *******************************************
100 CLS : K6= -1: SK1= 2: XDIR= 1: T9= TIMER 'XDIR: 1=FWD, 0=INVERSE
102 GOSUB 200 ' DO FORWARD ROW XFORMS
104 GOSUB 300 ' DO FORWARD COLUMN XFORMS
106 T9 = TIMER - T9 ' CHECK TIME
112 RETURN
' *******************************************
' *            INVERSE TRANSFORM          *
' *******************************************
150 CLS : K6 = 1: SK1 = 1: XDIR = 0: T9 = TIMER
152 GOSUB 300 ' RECONSTRUCT COLUMNS
153 GOSUB 200 ' RECONSTRUCT ROWS
155 T9 = TIMER - T9 ' GET TIME
160 RETURN
```

```
'   **************************************************
'   *                    PLOT DATA                   *
'   **************************************************
176 CLS : AMP1 = 0 ' FIND LARGEST MAGNITUDE IN ARRAY
178     FOR I = 0 TO Q - 1
180         FOR J = 0 TO Q - 1
182             IF XDIR = 0 THEN AMP = C(I, J): GOTO 186
184             AMP = SQR(C(I, J) ^ 2 + S(I, J) ^ 2)
186             IF AMP > AMP1 THEN AMP1 = AMP
188         NEXT J
190     NEXT I
192 MAG2 = -130 / AMP1 ' SET SCALE FACTOR
194 GOSUB 6000 ' PLOT 2-D DATA
196 LOCATE 1, 1: PRINT "TIME = "; T9
198 RETURN

'   **************************************************
'   *                   TRANSFORMS                   *
'   **************************************************
200 CLS : KRTST = 19
202 FOR KR = 0 TO Q1 ' XFORM 2D ARRAY BY ROWS
204 'IF XDIR = 1 THEN GOSUB 400
206 PRINT USING "###_ "; KR; ' PRINT ROW BEING XFORMED
208 IF KR = KRTST THEN PRINT : KRTST = KRTST + 20' END PRINT LINE

'   ************************************
'   * THE ROUTINE BELOW IS FOR A ROW   *
'   ************************************
210 FOR M = 0 TO N1: QT = 2 ^ (N - M)' DO N STAGES
212 QT2 = QT / 2: QT3 = QT2 - 1: KT = 0
214 FOR J = 0 TO Q1 STEP QT: KT2 = KT + 1' DO ALL FREQUENCY SETS
216 FOR I = 0 TO QT3: J1 = I + J: K = J1 + QT2' DO FREQUENCIES IN SET
    ' ROW BUTTERFLY
218 CTEMP = (C(KR,J1) + C(KR,K)*KC(KT) - K6*S(KR,K)*KS(KT))/SK1
220 STEMP = (S(KR,J1) + K6*C(KR,K)*KS(KT) + S(KR,K)*KC(KT))/SK1
222 CTEMP2 = (C(KR,J1) + C(KR,K)*KC(KT2) - K6*S(KR,K)* KS(KT2))/SK1
224 S(KR,K) = (S(KR,J1) + K6*C(KR,K)*KS(KT2) + S(KR,K)*KC(KT2))/SK1
226 C(KR, K) = CTEMP2: C(KR, J1) = CTEMP: S(KR, J1) = STEMP
228 NEXT I' ROTATE AND SUM NEXT PAIR OF COMPONENTS
230 KT = KT + 2
232 NEXT J' DO NEXT SET OF FREQUENCIES
234 NEXT M' DO NEXT STAGE
    ' BIT REVERSAL FOR ROW TRANSFORMS
236 FOR I = 1 TO Q1: INDX = 0
238 FOR J = 0 TO N1
240 IF I AND 2 ^ J THEN INDX = INDX + 2 ^ (N1 - J)
242 NEXT J
244 IF INDX>I THEN SWAP C(KR,I),C(KR,INDX): SWAP S(KR,I),S(KR,INDX)
246 NEXT I
248 'IF XDIR = 0 THEN GOSUB 400
250 NEXT KR
252 T9 = TIMER - T9: GOSUB 176' USE TO SHOW RESULTS OF ROW XFORMS
254 A$ = INKEY$: IF A$ = "" THEN 254
256 CLS : T9 = TIMER - T9: RETURN' ROW TRANSFORMS DONE
```

The 2-D FFT

```
'   ***************************************
'   * THE ROUTINE BELOW IS FOR COLUMNS    *
'   ***************************************
300 KRTST = 19
302 FOR KR = 0 TO Q1 ' XFORM 2D ARRAY BY COLUMNS
304 'IF XDIR = 1 THEN GOSUB 410
306 PRINT USING "###_ "; KR;
308 IF KR = KRTST THEN PRINT : KRTST = KRTST + 20
310 FOR M = 0 TO N1: QT = 2 ^ (N - M)
312 QT2 = QT / 2: QT3 = QT2 - 1: KT = 0
314 FOR J = 0 TO Q1 STEP QT: KT2 = KT + 1
316 FOR I = 0 TO QT3:  J1 = I + J: K = J1 + QT2
    'COLUMN BUTTERFLYS
318 CTEMP = (C(J1,KR) + C(K,KR)*KC(KT) - K6*S(K,KR)*KS(KT))/SK1
320 STEMP = (S(J1,KR) + K6*C(K,KR)*KS(KT) + S(K,KR)*KC(KT))/SK1
322 CTEMP2 = (C(J1,KR) + C(K,KR)*KC(KT2) - K6*S(K,KR)*KS(KT2))/SK1
324 S(K,KR) = (S(J1,KR) + K6*C(K,KR)*KS(KT2) + S(K,KR)*KC(KT2))/SK1
326 C(K, KR) = CTEMP2: C(J1, KR) = CTEMP: S(J1, KR) = STEMP
328 NEXT I
330 KT = KT + 2
332 NEXT J
334 NEXT M
'   ***************************************
'   * BIT REVERSAL FOR COLUMN TRANSFORMS  *
'   ***************************************
336 FOR I = 1 TO Q1: INDX = 0
338 FOR J = 0 TO N1
340 IF I AND 2 ^ J THEN INDX = INDX + 2 ^ (N1 - J)
342 NEXT J
344 IF INDX>I THEN SWAP C(I,KR),C(INDX,KR): SWAP S(I,KR),S(INDX,KR)
346 NEXT I
348 'IF XDIR = 0 THEN GOSUB 410
350 NEXT KR
352 IF K6 = 1 THEN XDIR = 1
354 T9 = TIMER - T9: GOSUB 176' USE TO SHOW RESULTS OF COLUMN XFORMS
356 IF K6 = 1 THEN XDIR = 0
358 A$ = INKEY$: IF A$ = "" THEN 358
360 CLS : T9 = TIMER - T9: RETURN' COLUMN TRANSFORMS DONE

'   ***************************************
'   *         MODIFY ROW SAMPLING         *
'   ***************************************
400 FOR I = 1 TO Q1 STEP 2
402 C(KR, I) = -C(KR, I): S(KR, I) = -S(KR, I)
404 NEXT I
406 RETURN

'   ***************************************
'   *       MODIFY COLUMN SAMPLING        *
'   ***************************************
410 FOR I = 1 TO Q1 STEP 2
412 C(I, KR) = -C(I, KR): S(I, KR) = -S(I, KR)
414 NEXT I
416 RETURN
```

```
'     **********************************
'     *      GENERATE FUNCTIONS        *
'     **********************************
5000 XDIR = 0
5001 CLS : PRINT : PRINT : PRINT "           FUNCTION MENU": PRINT
5002 PRINT " 1 = GENERATE SINC^2 FUNCTION   2 = SINC FUNCTION":PRINT
5009 PRINT " 9 = EXIT:": PRINT
5010 PRINT "            MAKE SELECTION";
5012 A$ = INKEY$: IF A$ = "" THEN 5012
5014 A = VAL(A$): ON A GOTO 5030, 5500
5016 IF A = 9 THEN RETURN
5018 GOTO 5000

'     **********************************
'     *        SINC^2 FUNCTION         *
'     **********************************
5030 CLS
5032 INPUT "WIDTH"; WDTH1 ' INPUT FINCTION SIZE
5034 IF WDTH1 = 0 THEN WDTH1 = 1 ' ZERO INVALID
5036 SKL1 = PI2 / WDTH1: MAG1 = Q ' CONSTANTS
5038 FOR I = 0 TO Q - 1 '
5040 YARG = SKL1 * (I - Q2): PRINT "*";
5042 FOR J = 0 TO Q - 1
5044 XARG = SKL1 * (J - Q2)
5046 IF YARG = 0 AND XARG = 0 THEN C(I, J) = MAG1: GOTO 5052
5048 ARG = SQR(XARG ^ 2 + YARG ^ 2)
5050 C(I, J) = MAG1 * (SIN(ARG) / ARG) ^ 2: S(I, J) = 0
5052 NEXT J
5054 NEXT I
5055 MAG2 = -130 / MAG1
5056 GOSUB 6000 ' PLOT FUNCTION
5058 PRINT "ANY KEY TO CONTINUE";
5060 A$ = INKEY$: IF A$ = "" THEN 5060 ' WAIT
5080 RETURN

5500 '    **********************************
'         *         SINC FUNCTION          *
'         **********************************
5502 CLS : MAG1 = Q: T0 = 1: T1 = 0
5504 INPUT "WIDTH"; WDTH1
5506 SKL1 = PI2 / WDTH1: MAG1 = Q
5508 FOR I = 0 TO Q - 1
5510 YARG = SKL1 * (I - Q2): PRINT "*";
5520 FOR J = 0 TO Q - 1
5522 XARG = SKL1 * (J - Q2)
5524 IF YARG = 0 AND XARG = 0 THEN C(I, J) = MAG1: GOTO 5530
5526 ARG = SQR(XARG ^ 2 + YARG ^ 2)
5528 C(I, J) = MAG1 * (SIN(ARG) / ARG): S(I, J) = 0
5530 NEXT J
5550 NEXT I
5590 GOSUB 176
5592 PRINT "ANY KEY TO CONTINUE";
5594 A$ = INKEY$: IF A$ = "" THEN 5594
5598 RETURN
```

The 2-D FFT

```
6000  '  *******************************
      '  *        PLOT DATA            *
      '  *******************************
6002  CLS ' CLEAR SCREEN AND SET SCALE FACTORS
6004  XCAL = 320 / Q: YCAL = 120 / Q: YDIS = 150: X0 = 15
6006  FOR I = 0 TO Q - 1 ' FOR ALL ROWS
6008  DISP = X0 + (Q - I) * 288 / Q ' DISPLACE ROWS FOR 3/4 VIEW
6010  PER = I / (2 * Q) ' CORRECT FOR PERSPECTIVE
6012  FOR J = 0 TO Q - 1 ' FOR EACH PIXEL IN ROW
6014  X11 = ((XCAL + PER)*J) + DISP: Y11 = ((YCAL + .3*PER)*I) + YDIS
6016  IF XDIR= 0 THEN AMP= C(I,J) ELSE AMP= SQR(C(I,J)^2+ S(I, J)^2)
6018  AMP = MAG2 * AMP
6020  LINE (X11, Y11 + AMP)-(X11, Y11)
6022  PRESET (X11, Y11 + AMP + 1)
6024  NEXT J ' NEXT PIXEL
6026  NEXT I ' NEXT ROW
6028  RETURN ' ALL DONE
      ' *************
9999  END: STOP
```

Displaying the 2-D function (lines 6000-28) must be discussed. We scale the X axis to cover 320 pixels for Q data points (6004) and Y = 120/Q. Instead of defining Y0 we call it YDIS = 150, and along with X0 we define DISP in line 6008. Note that DISP changes as we step down through the rows, moving slowly to the left. In line 6014 we calculate the actual location of the pixel as X11 and Y11. At 6018 we calculate the *intensity* of the function (AMP) depending on whether it is a frequency or time domain function (i.e., the Z axis component), and at line 6020 we plot this as a *vertical* line. This simple routine results in a "3-D" representation of the function. To make this routine work we must know the maximum amplitude of the function to be plotted (MAG2), and we take care of that requirement back in lines 176-198. In lines 178-190 we step through the data points and compare the magnitudes to a test value AMP1. If the data point exceeds AMP1 it becomes AMP1. In line 192 we use AMP1 to calculate MAG2. At line 194 we jump to the plot routine.

Lines 5000-5018 are the menu for the generate functions routines. We use only two functions here and they both have the same general form. You will recognize the argument (calculated as a radius in line 5048) is an *angular* function, and the sinc function squared is calculated in line 5050. We set the scale factor MAG2 = -130/MAG1 and jump directly to the plot routine.

The one major modification to this routine (i.e., quadrant correction) may be implemented by simply removing the apostrophe from in front of lines 204, 248, 304 and 348; however, you should always feel free to modify these routines in any way you desire.

APPENDIX 14.1

2-D FUNCTIONS

14.1 QUADRANT SWITCHING

While the *sampling modification* of the previous chapter gets the quadrants in the proper location, we may also solve the problem by simply switching the quadrants around. The code is shown below, which replaces the modify sampling routine at lines 400-434.

```
' *******************************************
' *         QUADRANT CORRECTION             *
' *******************************************
400 FOR I = 0 TO Q3
402   FOR J = 0 TO Q3
404     I2 = I + Q2: J2 = J + Q2
406     SWAP C(I2, J2), C(I, J): SWAP S(I2, J2), S(I, J)
410   NEXT J
412 NEXT I
420 FOR I = 0 TO Q3
422   FOR J = Q2 TO Q - 1
424     I2 = I + Q2: J2 = J - Q2
426     SWAP C(I2, J2), C(I, J): SWAP S(I2, J2), S(I, J)
430   NEXT J
432 NEXT I
434 RETURN
```

To solve the problem completely we must switch before taking the transform and *after* taking the transform—the shifts are identical in both cases. We must switch them before and after taking the *inverse transform*, of course—you might want to experiment with what happens if you only switch once (the *problem* becomes apparent in Ch. 16).

The first *function* we consider is the sinusoidal "bar" pattern.

```
5200 REM **********************************
     REM *         SINUSOIDAL BARS        *
     REM **********************************
5202 CLS : KRAD = PI / Q
5204 PRINT "INPUT HORIZ, VERT FREQUENCIES (MUST BE BETWEEN 1 AND"; Q2;
5206 INPUT FH, FV: IF FH < 0 OR FH > Q2 THEN 5204
5208 IF FV < 0 OR FV > Q2 THEN 5204
5210 HKRAD = KRAD * FH: VKRAD = KRAD * FV
5212 FOR I = 0 TO Q1
5214 FOR J = 0 TO Q1
5216 C(I, J) = SIN(VKRAD * J - HKRAD * I) ^ 2: S(I, J) = 0
5224 NEXT J
5226 NEXT I
5228 MAG2 = -130: GOSUB 6000' DISPLAY FUNCTION
5230 INPUT A$   ' WAIT USER INPUT
5232 RETURN
```

The 2-D Functions

This is another simple function. We input the frequencies of both the horizontal and vertical sinusoids (if you want horizontal bars only, input 0 for the vertical frequency, etc). We calculate the angular velocity for vertical and horizontal in line 5210, and lines 5212-26 set up loops to step through the data points. *J* designates the column; so, if the horizontal angular velocity is zero, we will generate the same values for each column (similar horizontal results for zero vertical angular velocity).

We may conveniently make the point here that a digitized image is a matrix of *real valued* numbers. Consequently, we could use the PFFFT for the row transforms, cutting the transform time almost in half! For a reasonably large picture this would be a significant gain; but, the results of these row transforms are *complex numbers*, and a conventional FFT is required for the columns. Nevertheless, we would only have half as many components to transform if we used a PFFFT for the rows, so the 2:1 speedup would be carried through the whole 2-D transform.

The connection with the sinusoidal bar generator is this: there are two ways to generate the 2-D sinusoids in line 5716—we may either *add* the x and y variables or we may *subtract* them (in the present case we subtract them). The difference in the final function is that the peaks and troughs of the sinusoids either run from *upper left* to *lower right* or from *lower left* to *upper right*. Both of these forms are necessary to draw all possible pictures of course. The difference in the frequency domain is that one of these forms shows up as *negative frequencies* in the 2-D transform, and that is why we *must* use a conventional, negative and positive frequency FFT in the column transforms. It's not especially difficult to construct a 2-D transform using the PFFFT for the horizontal transforms and would be a good exercise for anyone who plans to spend a lot of time working with images (i.e., 2-D data formats).

The second function we must generate is the sinc^2 function.

```
'    *********************************
'    *        SINC^2 FUNCTION        *
'    *********************************
5030 CLS : T1 = 0: T0 = 1
5032 INPUT "WIDTH"; WDTH1  ' INPUT FINCTION SIZE
5034 IF WDTH1 = 0 THEN WDTH1 = 1  ' ZERO INVALID
5036 SKL1 = PI2 / WDTH1: MAG1 = Q  ' CALC CONSTANTS
5038 FOR I = 0 TO Q - 1 '
5040 YARG = SKL1 * (I - Q2): PRINT "*";
5042 FOR J = 0 TO Q - 1
5044 XARG = SKL1 * (J - Q2)
5046 IF YARG = 0 AND XARG = 0 THEN C(I, J) = MAG1: GOTO 5052
```

```
5048 ARG = SQR(XARG ^ 2 + YARG ^ 2)
5050 C(I, J) = MAG1 * (SIN(ARG) / ARG) ^ 2: S(I, J) = 0
5052 NEXT J
5054 NEXT I
5055 MAG2 = -130 / MAG1
5056 GOSUB 6000 ' PLOT FUNCTION
5058 INPUT A$ ' WAIT
5060 RETURN
```

We know the sinc function is simply the sine of the argument divided by the argument, but the argument in this case is the radial distance from the center of the function. To draw this function in the center of the matrix we make the zero at point *(Q/2, Q/2)*. We enter a desired "width" for our function at line 5032. Line 5036 then calculates a scale factor for that width. At line 5038 we set up a loop which counts the rows of the matrix, and at 5040 calculate the Y axis argument as SKL1*(I-Q2) [thereby making the zero point at the middle row]. Line 5042 sets up a nested loop to count the columns and then we calcuate an X axis argument which will be zero half-way across the columns. We must then check for the 0,0 point to avoid division by zero. At line 5048 we calculate the *radial argument* for the sinc^2 function, and at line 5050 calculate the value for this *x, y* position.

We next consider the Bessel function routine. The equations for this function may be found in any good reference book (e.g., the *CRC Handbook of Chemistry and Physics*, CRC Press). There are a great many ways to implement a routine to generate these functions and, except for the following comments, we will not go into detail on this particular routine. The routine is presented below.

```
     ' *********************************
     ' *       BESSEL FUNCTION         *
     ' *********************************
5600 CLS : DEFDBL D-K
5604 INPUT "WIDTH"; WDTH1
5606 IF WDTH1 < 1 THEN 5604 ' MINIMUM WIDTH
5608 SKL1 = PI / (3.6 * WDTH1 * Q / 64)
5610 FOR I = 0 TO Q - 1
5612 YARG = SKL1 * (I - Q2): PRINT "*";
5614 FOR J = 0 TO Q - 1
5616 XARG = SKL1 * (J - Q2)
5618 KARG = SQR(XARG ^ 2 + YARG ^ 2)
5620 KA = 1: KB = 1: DAT1 = 1: KTGL = 1
5622 FOR K = 2 TO 900 STEP 2
5624 KTGL = -1 * KTGL
5626 KA = KA * K: KB = KB * (K + 2): DENOM = KA * KB
5628 DAT2 = KTGL * (WDTH1 ^ (K / 2) * KARG ^ K / DENOM)
5630 IF ABS(DAT2) < ABS(DAT1) * 1E-10 THEN 5640
5632 DAT1 = DAT1 + DAT2
```

The 2-D Functions

```
5634  '  PRINT DAT1,
5636 NEXT K
5638 PRINT "#"
5640 C(I, J) = DAT1: S(I, J) = 0
5642 NEXT J
5644 NEXT I
5646 GOSUB 176
5648 INPUT A$
5650 RETURN
```

The *correct* aperture function, however, may be generated via a modification of the sinusoidal bar routine.

```
     REM ********************************
     REM *     BESSEL FUNCTION II       *
     REM ********************************
5800 CLS: MAG1 = Q: KR = PI2 / Q: MAG2 = -130
5802 FOR I= 0 TO Q: FOR J= 0 TO Q: C(I,J)= 0: S(I,J) = 0: NEXT: NEXT
5806 INPUT "WIDTH"; WDH
5808 FOR A = 0 TO WDH: IF A = 0 THEN M2 = 0.5 ELSE M2 = 1.0
5810 FOR B= 0 TO SQR(WDH^2 - A^2): IF B= 0 THEN M1 = M2/2 ELSE M1= M2
5812 FOR I = 0 TO Q1: I2 = A * (I - Q2)
5814 FOR J = 0 TO Q1: J2 = B * (J - Q2)
5816 C(I,J) = C(I,J) + M1*(COS(KR*(I2 - J2)) + COS(KR*(I2 + J2)))
5818 NEXT J: NEXT I
5822 NEXT B: NEXT A
5828 GOSUB 176
5830 INPUT A$
5832 RETURN
```

Sinusoidal bar patterns transform as single harmonics in the frequency domain; therefore, we may place components at any position. To generate the correct aperture function in the frequency domain we need only sum the sinusoidal bars required to cover the desired aperture.

We ask for the desired "width" in line 5806, and in line 5808 we set up a loop to step A from zero to this width. A will specify the frequency of the sinusoids down the rows by stepping through the harmonic frequencies from zero to WDH, so we must then step B through all the frequencies that are to be generated within that row. For example, when $A = 0$, B must step through every frequency up to the WDH specification; however, when $A =$ WDH, B will only take on the value of zero. That is, B must take on all values to generate a circular pattern in the frequency domain, which is accomplished by solving for the x, y components of a radius: $B = \sqrt{WDH^2 - A^2}$. So, for each of these frequencies we must perform the generate *sinusoidal bars* routine (lines 5812-5818). You will notice that we do *not* square the sinusoids in this routine [you might want to try that variation in your experiments]. You should note the difference between Fig. 14.14 and Fig. 14.11 in the text.

Finally, we can generate the proper image domain function by using the inverse of the above process. We generate a circular function in the frequency domain and inverse transform (the Bessel III routine).

```
5300  '   **********************************
      '   *           CIRC FUNCTION        *
      '   **********************************
5302  CLS : MAG1 = Q
5304  INPUT "DIAMETER"; DIA1
5306  INPUT "CENTERED ON (X,Y)"; CNTRX, CNTRY
5308  SKL1 = Q / DIA1: MAG1 = Q
5310  FOR I = 0 TO Q - 1
5312  YARG = I - CNTRY: PRINT "*";
5314  FOR J = 0 TO Q - 1
5316  XARG = J - CNTRX
5318  C(I, J) = 0:
5320  ARG = SQR(XARG ^ 2 + YARG ^ 2)
5322  IF ARG <= DIA1 THEN C(I, J) = MAG1: S(I, J) = 0
5324  NEXT J
5326  NEXT I
5328  GOSUB 176
5330  INPUT A$
5332  RETURN
```

This routine is so simple it doesn't require an explanation, but it should be pointed out that for the aperture function, the function should be centered at Q/2, Q/2. [Note: If you construct this function in the *frequency domain* as it is presented here, when you reconstruct the image domain, (without the benefit of quadrant re-arrangement) the reconstructed function will be split into quarters as was the first 2-D frequency domain function. A little thought will make it apparent why.]

We will not discuss the *star* or *impulse* function as it simply places a data point at some desired location in the image.

In regard to the Bessel function, we should note that Professor Gaskill[1] defines a "sombrero function" as a Bessel function of the first kind of the *first order* (by dividing its argument analogous to the construction of the sinc function). This is the function generated by a star (or other point source) when imaged through an optical system with a circular aperture.

---

[1] In *Linear Systems, Fourier Transforms, and Optics*, Jack D. Gaskill, John Wiley & Sons. So far as I know this is the standard graduate-level text in this field.

APPENDIX 15.1

FOURIER OPTICS ROUTINES

```
5400 '***********************************
     '*        DIFFRACTION TEST         *
     '***********************************
5402 CLS
5404 INPUT "INPUT SLIT WIDTH, LENGTH (I.E. TO 2,20)"; DELX, DELY
5406 MAG1 = Q ^ 2: ANM = MAG1 * ANP: AN = 0
5408 FOR I = 0 TO Q - 1
5410 FOR J = 0 TO Q - 1
5412 C(I, J) = AN: S(I, J) = AN
5414 NEXT J
5415 NEXT I
5416 IF DELX = 0 THEN 5450 'HEXAGON APERTURE
5417 IF DELY = 0 THEN 5468 'OCTAGON APERTURE
5418 IF DELX > Q2 OR DELY > Q2 THEN 5425
5420 FOR I = -DELX TO (DELX - 1)
5421 FOR J = -DELY TO (DELY - 1)
5422 C(Q2 - I, Q2 + J) = MAG1
5423 NEXT J
5424 NEXT I
5425 ' C(0, 0) = MAG1
5426 MAG2 = -120 / Q ^ 2
5428 GOSUB 176 ' DISPLAY HANDIWORK
5430 INPUT "C/R TO CONTINUE"; A$
5432 RETURN
' *******   HEX APERTURE    *******
5450     FOR I = -6 TO 6: A = -CINT(ABS(I) * .6285 - 6.8667)
5452     FOR J = -(1 + A) TO A
5454     C(Q2 - I, Q2 + J) = MAG1
5456     NEXT J
5458     NEXT I
5464     GOTO 5426
' **********   OCTAGON APERTURE   *******
5468 FOR I = 4 TO 8
5470 FOR J = -I TO I - 1: C(Q2 - 12 + I, Q2 + J) = MAG1: NEXT J
5472 NEXT I
5474 FOR I = 9 TO 14
5476 FOR J = -8 TO 7: C(Q2 - 12 + I, Q2 + J) = MAG1: NEXT J
5478 NEXT I
5480 FOR I = 8 TO 4 STEP -1
5482 FOR J = -I TO I - 1: C(Q2 + 11 - I, Q2 + J) = MAG1: NEXT J
5484 NEXT I
5486 GOTO 5426
```

This routine generates a slit of user selected dimensions (lines 5420-5424); however, if a zero is entered for the x dimension a small hexagon shaped aperture will be created (lines 5450-5464) and if a zero is entered for the y dimension an octagon shaped aperture will be created (lines 5468-5486). You might want to look closely at how these are generated—especially the hex aperture.

```
'   ***********************************************************
'   *                    PRINT OUTPUT
'   ***********************************************************
500 CLS : PRINT "ONLY DATA ABOUT THE 0,0 POINT WILL BE PRINTED"
'   ***********************************************************
540 CLS : PRINT "     ONLY THE PHASES OF THE HARMONICS ARE PRINTED":
PRINT ' PRINT SCREEN
542 FOR I = Q2 - 3 TO Q2 + 3
544 FOR ROLOC = Q2 - 3 TO Q2 + 3
546 Y2 = C(ROLOC, I): X2 = S(ROLOC, I): MAG = SQR(Y2 ^ 2 + X2 ^ 2)
548 ANG = 180 / PI * ATN(X2 / Y2): IF Y2 < 0 THEN ANG = ANG + 180
550 PRINT USING "###.###_   "; ANG;
552 NEXT ROLOC
554 PRINT : NEXT I
556 INPUT A$
558 RETURN
```

This Print Routine is the one used to generate the tables of *p.* 180 (Chapter 15, section 15.6), and will display only the angle of the transformed harmonics (for ± 3 rows and ± 3 columns about the 0, 0 location).

APPENDIX 16.1

IMAGE ENHANCEMENT

```
'   *******************************************
'   *******  2DFFT 16.00 - 2D XFORM   *******
'   *******************************************
10 SCREEN 9, 1, 1: COLOR 15, 1: CLS
14 INPUT "SELECT ARRAY SIZE AS 2^N. N ="; N
16 N1 = N - 1: Q = 2 ^ N: Q1 = Q - 1
18 ' $DYNAMIC
20 DIM C(Q, Q), S(Q, Q), KC(Q), KS(Q), DAC(Q, Q), DAS(Q, Q)
22 TYPE Pixl
      PIX AS INTEGER
   END TYPE
28 DIM PIXREC AS Pixl
30 Q2 = Q / 2: Q3 = Q2 - 1: Q4 = Q / 4: Q5 = Q4 - 1: Q8 = Q / 8
32 PI = 3.141592653589793#: PI2 = 2 * PI: K1 = PI2 / Q: CLVK = 1
   ' **** TWIDDLE FACTOR TABLE GENERATION ****
40 FOR I = 0 TO Q: KC(I) = COS(K1 * I): KS(I) = SIN(K1 * I)
42 IF ABS(KC(I)) < .0000005 THEN KC(I) = 0 ' CLEANUP TABLE
44 IF ABS(KS(I)) < .0000005 THEN KS(I) = 0
46 NEXT I
48 FOR I = 1 TO Q1: INDX = 0
50 FOR J = 0 TO N1
52 IF I AND 2 ^ J THEN INDX = INDX + 2 ^ (N1 - J)
54 NEXT J
56 IF INDX > I THEN SWAP KC(I), KC(INDX): SWAP KS(I), KS(INDX)
58 NEXT I
   ' ***************
70 CLS : PRINT : PRINT : PRINT "               MAIN MENU": PRINT
72 PRINT " 1 = LOAD .BMP FILE": PRINT
74 PRINT " 2 = TRANSFORM FUNCTION": PRINT
76 PRINT " 3 = INVERSE TRANSFORM ": PRINT
78 PRINT " 4 = MODIFY TRANSFORM           ": PRINT
80 PRINT " 5 = SAVE FILE              ": PRINT
84 PRINT " 6 = GENERATE FUNCTIONS       ": PRINT
86 PRINT " 7 = MODIFY SPACE DOMAIN FUNCTION": PRINT
88 PRINT " 9 = EXIT                  ": PRINT : PRINT
90 PRINT "          MAKE SELECTION";
92 A$ = INKEY$: IF A$ = "" THEN 92
94 A = VAL(A$): ON A GOSUB 4000, 100, 150, 3000, 4100, 5000, 2400
96 IF A = 9 THEN 9999
98 GOTO 70
   ' ********** XFORM FUNCTION GOES HERE **************
2400 ' *** MODIFY SPACE DOMAIN FUNCTION ***
2402 CLS : PRINT : PRINT SPC(15); "MENU"
2404 PRINT : PRINT "1 = CHANGE GREY SCALE RANGE"
2406 PRINT : PRINT "2 = MODIFY CONTRAST"
2408 A$ = INKEY$: IF A$ = "" THEN 2408
2410 ON VAL(A$) GOSUB 2420, 2460
2412 RETURN
' *** LIMIT GREY SCALE RESOLUTION ***
2420 CLS
2422 A = 1: B = 0 ' FIRST FIND RANGE OF DATA
2424    FOR I = 0 TO Q - 1
2426       FOR J = 0 TO Q - 1
```

```
2428                IF C(I, J) > A THEN A = C(I, J)
2430                IF C(I, J) < B THEN B = C(I, J)
2432            NEXT J
2434        NEXT I
2436 IF B > 0 THEN B = 0 ' IF > 0 ELIMINATE FROM CALCS.
2438 INPUT "NUMBER OF BITS GREY SCALE (LESS THAN 16)"; GK
2440 GK = 2 ^ GK
2442 A = (A - B) / GK
2444     FOR I = 0 TO Q - 1
2446         FOR J = 0 TO Q - 1
2448             C(I, J) = INT((C(I, J) - B) / A)
2450         NEXT J
2452     NEXT I
2454 PRINT A, B' OKAY, NOW WE HAVE A NEW DYNAMIC RANGE
2456 RETURN
2460 ' *** CHANGE CONTRAST ***
2462 AMAX = 0: AMIN = 256
2464 FOR I = 0 TO Q - 1
2466 FOR J = 0 TO Q - 1
2468 A = C(I, J): IF A > AMAX THEN AMAX = A
2470 IF A < AMIN THEN AMIN = A
2472 NEXT J
2474 NEXT I
2475 C = 0: PRINT "AMAX ="; AMAX; " AND AMIN ="; AMIN
2476 PRINT "HOLD BLACK OR WHITE CONSTANT (B/W)?"
2478 A$ = INKEY$: IF A$ = "" THEN 2478
2480 IF A$ = "B" OR A$ = "b" THEN C = 1: GOTO 2484
2482 IF A$ <> "W" AND A$ <> "w" THEN 2478
2484 PRINT : INPUT "INPUT MULTIPLICATION FACTOR"; MK
2486 FOR I = 0 TO Q - 1
2488 FOR J = 0 TO Q - 1
2490 IF C = 0 THEN C(I, J) = AMAX - (MK * (AMAX - C(I, J)))
2492 IF C = 1 THEN C(I, J) = (MK * (C(I, J) - AMIN)) + AMIN
2494 NEXT J
2496 NEXT I
2498 GOSUB 6000 ' DISPLAY RESULTS
2500 INPUT A$ ' WAIT
2502 RETURN
' THE ROUTINES BELOW REQUIRE DIMENSIONING ARRAY STD(Q, Q)
' ********** SAVE PICTURE **********
'     FOR I = 0 TO Q - 1
'         FOR J = 0 TO Q - 1
'             STD(I, J) = C(I, J)
'         NEXT J
'     NEXT I
RETURN
' ********** RECALL PICTURE **********
'     FOR I = 0 TO Q - 1
'         FOR J = 0 TO Q - 1
'             C(I, J) = STD(I, J): S(I, J) = 0
'         NEXT J
'     NEXT I
RETURN
        ' **********************************
        ' *        MODIFY TRANSFORM        *
```

Image Enhancement

```
     ' *********************************
3000 CLS : PRINT : PRINT "MODIFY SPECTRUM MENU": PRINT : PRINT
3002 PRINT " 1 - MULTIPLY WITH STORED SPECTRUM"
3004 PRINT " 2 - DIVIDE CURRENT/STORED SPECTRUMS"
3006 PRINT " 3 - FILTER SPECTRUM"
3008 PRINT " 4 - TRUNCATE SPECTRUM"
3010 PRINT " 5 - SAVE SPECTRUM"
3014 PRINT " 9 - EXIT": PRINT : PRINT
3018 PRINT "    MAKE SELECTION"
3020 A$ = INKEY$: IF A$ = "" THEN 3020
3022 IF ASC(A$) < 49 OR ASC(A$) > 57 THEN 3000
3024 A = VAL(A$): ON A GOSUB 3200, 3300, 3100, 3500, 3400
3026 IF A = 9 THEN A = 8
3028 RETURN
     ' *********************************
     ' *    GAUSSIAN FILTER
     ' *********************************
3100 INPUT "CUTOFF FREQ. "; FCO
3102 FOR I = -Q2 TO Q2
3104    FOR J = -Q2 TO Q2
3106       FI = SQR(I^2 + J^2): ATTN = (10^(-.3/2*(FI/FCO)^3))
3108       I2 = Q2 + I: J2 = Q2 + J
3110       C(I2, J2) = C(I2, J2)*ATTN: S(I2, J2) = S(I2, J2)*ATTN
3112    NEXT J
3114 NEXT I
3116 GOSUB 6000 ' DISPLAY RESULTS
3118 INPUT A$ ' WAIT
3120 RETURN
3200 ' **** MULTIPLY FREQUENCY FUNCTIONS ****
3202 FOR I = 0 TO Q - 1
3204    FOR J = 0 TO Q - 1
3206       CIJ = C(I, J): SIJ = S(I, J)
3208       CDAT = DAC(I, J): SDAT = DAS(I, J)
3210       C(I,J)=(CIJ*CDAT)-(SIJ*SDAT):S(I,J)=(CIJ*SDAT)+(SIJ*CDAT)
3212    NEXT J
3214 NEXT I
3216 GOSUB 6000 ' DISPLAY RESULTS
3228 INPUT A$ ' WAIT
3230 RETURN
3300 CLS ' **** DIVIDE CURRENT/STORED SPECTRUM ****
3302 PRINT SPC(5); "C = DIVIDE CURRENT FUNCTION BY STORED FUNCTION"
3304 PRINT SPC(5); "S = DIVIDE STORED FUNCTION BY CURRENT FUNCTION"
3308 A$ = INKEY$: IF A$ = "" THEN 3308
3310 CSFLG = 1: IF A$ = "S" OR A$ = "s" THEN CSFLG = 2: GOTO 3314
3312 IF A$ <> "C" AND A$ <> "c" THEN 3308
3314 FOR I = 0 TO Q - 1
3316 FOR J = 0 TO Q - 1
3318 CIJ = C(I, J): SIJ = S(I, J): C(I, J) = 0: S(I, J) = 0
3320 RIJ = SQR(CIJ ^ 2 + SIJ ^ 2)
3322 ' IF RIJ < .0000001 THEN 3348 ' SKIP IF DATA TOO SMALL ?
3324 IF CIJ = 0 THEN PHIJ = PI / 2: GOTO 3328 ' CAN'T DIVIDE BY ZERO
3326 PHIJ = ATN(SIJ / CIJ) ' FIND ANGLE
3328 IF CIJ < 0 THEN PHIJ = PHIJ + PI ' CORRECT FOR NEG X AXIS
3330 CDAT = DAC(I, J): SDAT = DAS(I, J) ' GET STORED DATA
3332 RDAT = SQR(CDAT ^ 2 + SDAT ^ 2) ' FIND MAG
```

```
3334 IF CDAT = 0 THEN PHDAT = PI/2: GOTO 3338 'CAN'T DIVIDE BY ZERO
3336 PHDAT = ATN(SDAT / CDAT) ' GET ANGLE
3338 IF CDAT < 0 THEN PHDAT = PHDAT + PI ' CORRECT FOR NEG X  AXIS
    ' *** DIVIDE (FINALLY) ***
3340 IF CSFLG = 1 THEN RQOT = RIJ/RDAT: PHQOT = PHIJ-PHDAT: GOTO 3344
3342 RQOT = RDAT / RIJ: PHQOT = PHDAT - PHIJ
3344 C(I, J) = COS(PHQOT) * RQOT ' CONVERT TO X,Y
3346 S(I, J) = SIN(PHQOT) * RQOT
3348 NEXT J
3350 NEXT I
3352 GOSUB 160 ' DISPLAY RESULTS
3354 INPUT A$ ' WAIT FOR USER
3390 RETURN
3400 ' ****  TEMP SAVE SPECTRUM IN DAX(I,J)  ****
3402 AMPN = 1
3404 FOR I = 0 TO Q - 1
3406    FOR J = 0 TO Q - 1
3408        DAC(I, J) = C(I, J): DAS(I, J) = S(I, J)
3410    NEXT J
3412 NEXT I
3414 RETURN
' ****  TRUNCATE SPECTRUM  ****
3500 ATTN = 0
3502 INPUT "TRUNCATE SPECTRUM ABOVE HARMONIC # "; HARX
3504 FOR I = 0 TO Q2
3506    FOR J = 0 TO Q2
3508        IF SQR(I ^ 2 + J ^ 2) < HARX THEN 3518
3510        GOSUB 3900
3518    NEXT J
3520 NEXT I
3522 GOSUB 6000 ' DISPLAY RESULTS
3524 INPUT A$ ' WAIT
3526 RETURN
3900 ' ****  MODIFY HARMONICS  ****
3902 I2 = Q2 + I: J2 = Q2 + J: I3 = Q2 - I: J3 = Q2 - J
3904 C(I2, J2) = C(I2, J2) * ATTN: S(I2, J2) = S(I2, J2) * ATTN
3906 C(I2, J3) = C(I2, J3) * ATTN: S(I2, J3) = S(I2, J3) * ATTN
3908 C(I3, J3) = C(I3, J3) * ATTN: S(I3, J3) = S(I3, J3) * ATTN
3910 C(I3, J2) = C(I3, J2) * ATTN: S(I3, J2) = S(I3, J2) * ATTN
3912 RETURN
       ' *********************************
       ' *         LOAD .BMP FILE        *
       ' *********************************
4000 CLS : PRINT : PRINT : PRINT
4002 DMAX = 20: XDIR = 0: PATH$ = "A:\"
4004 REDIM IMAGE%(DMAX) ' *.BMP HEADER ARRAY
4006 INPUT "FILE NAME (.BMP FILES ONLY)"; NM$
4008 OPEN PATH$ + NM$ + ".BMP" FOR RANDOM AS #3 LEN = LEN(PIXREC)
4010 PRINT " LOF = "; LOF(3)
4012 FOR J = 1 TO 20 ' READ HEADER
4014 GET #3, J, PIXREC
4016 IMAG% = PIXREC.PIX
4018 IMAGE%(J) = IMAG%
4024 NEXT J
4026 IF IMAGE%(1) = 19778 AND IMAGE%(10) = Q THEN 4040'CHK.BMP& SIZE
```

Image Enhancement

```
4030 INPUT " THIS IS NOT A LEGAL .BMP FILE - ENTER TO CONTINUE"; A$
4032 CLOSE 3: GOTO 4090 ' CLOSE FILE & EXIT
4040 STRT = IMAGE%(6) / 2 + 1 ' GET DATA START LOCATION
4044 LINX = IMAGE%(10) ' NO. LINES
4046 COLX = IMAGE%(12) ' NO. COLS.
     ' LOAD FILE INTO DATA ARRAY
4048 FOR I = 0 TO Q - 1: KL1 = STRT + (LINX*I/2) 'DATA LINE ADDRESS
4050 FOR J = 0 TO Q3 STEP 1: KL = KL1 + J: J2 = 2*J ' DATA ADDRESS
4052 GET #3, KL, PIXREC
4054 IMAG% = PIXREC.PIX
4056 C(I, J2) = IMAG% AND 255: C(I, J2 + 1) = (IMAG% AND 65280) / 256
4058 S(I, J2) = 0: S(I, J2 + 1) = 0
4060 NEXT J
4062 NEXT I
4064 CLOSE 3
4066 MAG2 = -.025 ' SCALE FACTOR
4068 GOSUB 6000 ' DISPLAY IMAGE
4074 INPUT "C/R WHEN READY"; A$ ' WAIT FOR USER
4090 RETURN
     ' *********************************
     ' *       SAVE *.BMP FILE         *
     ' *********************************
4100 CLS : PRINT : PRINT
4102 IF XDIR = 1 THEN GOSUB 50 ' TRANSFORM TO SPACE DOMAIN DATA
4104 A = 1: B = 0 ' FIRST BRING DATA BACK INTO RANGE, A=MAX, B=MIN
4106 FOR I = 0 TO Q - 1
4108    FOR J = 0 TO Q - 1
4110       IF C(I, J) > A THEN A = C(I, J)
4112       IF C(I, J) < B THEN B = C(I, J)
4114    NEXT J
4116 NEXT I
4118 IF B > 0 THEN B = 0 ' IF > 0 ELIMINATE FROM CALCULATIONS
4120 A = (A - B) / 255 ' FIND 0-255 SCALE FACTOR
4122 FOR I = 0 TO Q - 1 ' CONVERT TO 8 BIT DATA
4124    FOR J = 0 TO Q - 1
4126       C(I, J) = INT((C(I, J) - B) / A)
4128    NEXT J
4130 NEXT I
4132 FOR I = 0 TO 1: C(Q, I) = 0: C(I, Q) = 0: NEXT I
4134 PTH1$ = "C:\WINDOWS\D": PTH2$ = "C:\WINDOWS\": NM2$ = NM$
4136 IF LEN(NM2$) > 7 THEN NM2$ = LEFT$(NM2$, 7)
4138 OPEN PTH1$ + NM2$ + ".BMP" FOR RANDOM AS #3 LEN = LEN(PIXREC)
4140 OPEN PTH2$ + NM$ + ".BMP" FOR RANDOM AS #4 LEN = LEN(PIXREC)
4142 FOR I = 1 TO STRT - 1
4144 GET #4, I, PIXREC
4146 PUT #3, I, PIXREC
4148 NEXT I
4150 CLOSE #4
4152 PRINT "EXPAND FILE (Y/N)";
4154 A$ = INKEY$: IF A$ = "" THEN 4154
4156 IF A$ = "n" OR A$ = "N" THEN 4300
4158 PIXREC.PIX = INT(2 * Q)
4160 PUT #3, 10, PIXREC
4162 PUT #3, 12, PIXREC
4164 FOR I = 0 TO Q - 1: KS1 = (2 * I * Q) + STRT
```

```
4166 FOR J = 0 TO Q - 1: KS = KS1 + (J)
4168 PIXEL1 = C(I, J): PIXEL2 = (C(I, J + 1) + C(I, J)) / 2
4170 PIXEL = INT(PIXEL1) + 256 * INT(PIXEL2)
4172 IF PIXEL > 32767 THEN PIXEL = PIXEL - 65536
4174 PIXREC.PIX = INT(PIXEL)
4176 PUT #3, KS, PIXREC
4178 NEXT J
4180 KS1 = KS1 + Q
4182 FOR J = 0 TO Q - 1: KS = KS1 + J
4184 PIXEL1 = (C(I,J) + C(I + 1,J))/2
4185 PIXEL2 = (C(I,J) + C(I + 1,J) + C(I,J + 1) + C(I + 1,J + 1))/4
4186 PIXEL = INT(PIXEL1) + 256 * INT(PIXEL2)
4188 IF PIXEL > 32767 THEN PIXEL = PIXEL - 65536
4190 PIXREC.PIX = INT(PIXEL)
4192 PUT #3, KS, PIXREC
4194 NEXT J
4196 NEXT I
4198 CLOSE #3
4200 RETURN
4300 ' STORE UNEXPANDED PICTURE
4302 FOR I = 0 TO Q - 1: KS1 = (I*Q/2) + STRT
4304 FOR J = 0 TO Q - 1 STEP 2: KS = KS1 + (J/2)
4306 PIXEL1 = C(I,J): PIXEL2 = C(I,J + 1)
4308 PIXEL = INT(PIXEL1) + 256*INT(PIXEL2)
4310 IF PIXEL > 32767 THEN PIXEL = PIXEL - 65536
4312 PIXREC.PIX = INT(PIXEL)
4314 PUT #3, KS, PIXREC
4316 NEXT J
4318 NEXT I
4320 CLOSE #3
4322 RETURN
     ' *********************************
     ' *        GENERATE FUNCTIONS     *
     ' *********************************
5000 XDIR = 0
5001 CLS : PRINT : PRINT : PRINT "           FUNCTION MENU": PRINT
5002 PRINT " 1 = BESSEL FUNCTION           2 = GAUSSAIN": PRINT
5009 PRINT " 9 = EXIT:": PRINT
5010 PRINT "             MAKE SELECTION";
5012 A$ = INKEY$: IF A$ = "" THEN 5012
5014 A = VAL(A$): ON A GOTO 5600, 5700
5016 IF A = 9 THEN RETURN
5018 GOTO 5000
     ' ***         BESSEL FUNCTION       ***
5600 CLS : DEFDBL D-K
5604 INPUT "WIDTH"; WDTH1
5606 IF WDTH1 < 1 THEN 5604 ' MINIMUM WIDTH
5608 SKL1 = PI/(3.6*WDTH1*Q/64)
5610 FOR I = 0 TO Q - 1
5612 YARG = SKL1*(I - Q2): PRINT "*";
5614 FOR J = 0 TO Q - 1
5616 XARG = SKL1*(J - Q2)
5618 KARG = SQR(XARG^2 + YARG^2)
5620 KA = 1: KB = 1: DAT1 = 1: KTGL = 1
5622 FOR K = 2 TO 900 STEP 2
```

Image Enhancement

```
5624 KTGL = -1 * KTGL
5626 KA = KA * K: KB = KB * (K + 2): DENOM = KA * KB
5628 DAT2 = KTGL * (WDTH1 ^ (K / 2) * KARG ^ K / DENOM)
5630 IF ABS(DAT2) < ABS(DAT1) * 1E-10 THEN 5640
5632 DAT1 = DAT1 + DAT2
5636 NEXT K
5638 PRINT "#"
5640 C(I, J) = DAT1: S(I, J) = 0
5642 NEXT J
5644 NEXT I
5646 GOSUB 176
5648 INPUT A$
5650 RETURN
' ***           GAUSSIAN FUNCTION            ***
5700 CLS : MAG1 = Q / PI
5702 FOR I = 0 TO Q - 1: FOR J = 0 TO Q - 1
5704 C(I, J) = 0: S(I, J) = 0
5706 NEXT J: NEXT I
5708 INPUT "DIAMETER, CONSTANT"; DIA1, GCST
5710 MAG1 = (Q / DIA1) ^ 2 / PI
5712 FOR I = 0 TO Q2 - 1: I2 = I ' - Q2
5714 FOR J = 0 TO Q2 - 1: J2 = J ' - Q2
5716 R0 = SQR(I2^2 + J2^2): MAG11 = MAG1 * EXP(-(R0 / DIA1)^2)
5718 C(Q2 + I, Q2 + J) = MAG11: C(Q2 + I, Q2 - J) = MAG11
5720 C(Q2 - I, Q2 + J) = MAG11: C(Q2 - I, Q2 - J) = MAG11
5722 NEXT J
5724 NEXT I
5726 C(Q2, Q2) = C(Q2, Q2) + MAG1 * GCST
5728 XDIR = 1
5730 GOSUB 176
5732 INPUT A$
5734 RETURN
6000 ' ********************************
     ' *          PLOT DATA           *
     ' ********************************
6002 CLS ' CLEAR SCREEN AND SET SCALE FACTORS
6004 XCAL = 320 / Q: YCAL = 120 / Q: YDIS = 150
6006 FOR I = 0 TO Q1 ' FOR ALL ROWS
6008 DISP = (Q - I) * 288 / Q ' DISPLACE ROWS FOR 3/4 VIEW
6010 PER = I / (2 * Q) ' CORRECT FOR PERSPECTIVE
6012 FOR J = 0 TO Q1 ' FOR EACH PIXEL IN ROW
6014 X11 = ((XCAL + PER)*J) + DISP: Y11 = ((YCAL + .3*PER)*I) + YDIS
6016 IF XDIR= 0 THEN AMP= C(I,J) ELSE AMP= SQR(C(I,J)^2 + S(I,J)^2)
6018 AMP = MAG2 * AMP
6020 LINE (X11, Y11 + AMP)-(X11, Y11)
6022 PRESET (X11, Y11 + AMP + 1)
6024 NEXT J ' NEXT PIXEL
6026 NEXT I ' NEXT ROW
6028 RETURN ' ALL DONE
     ' **************
9999 END: STOP
```

We have included a 2-D Gaussian function (lines 5700-5734). This is just another radially symmetrical function equal to *EXP(x^2)*, and

Appendix 16.1

is similar to function routines of Appendix 14.1.

Okay, first of all, we will work with the *.BMP file format. As a non-professional you will find *all* of the file formats are a real pain. Pictures take incredible amounts of storage (easily exceeding several mega-bytes); consequently, most file formats use a *compression* scheme (and we don't want to get into that!). The *.BMP format usually stores data as a straightforward matrix of the pixels, so we only need to know enough about the file *header* to get data in and out of our array (after that you're on your own). There are actually two headers—the *.BMP header is 14 bytes and starts with ASCII "BM" (19778 in decimal). The next *double word* (i.e., 4 bytes) are the file length. The last double word is the *image data offset* which tells how far into the file the actual picture starts. Next comes the *information header*, with a double word for header size (40 bytes for our purposes). The next double word is *image width* (in pixels), followed by another double word specifying image height, and that's all we need to know about *that*. Let's look at loading a *.BMP file.

In lines 22-28 we use the TYPE/Pixel statement to define PIX as an integer (i.e., a double byte), then dimension PIXREC as Pixel. The reason for this obscure little ritual becomes apparent in the routine to load the file, which starts at line 4000. At 4002 we set DMAX equal to 20 (which is the number of words from the header we intend to read) and initialize things. In 4004 we redimension IMAGE%(DMAX) which we will use to save the header info (you recall the % sign defines the array as integer words). We get the file name and open file #3 in the Random Access mode. The statement LOF(3) returns the length of the file. We read the file header (lines 4012-24) and then check the first word to see if this is a *.BMP file, and also check (line 4026) to see if the tenth word is equal to the size of the array we have defined (i.e., the number of pixels along the side equals Q). If either of these conditions are not met we close the file and exit (lines 4030-32). We next get the image offset (line 4040) and generate the starting address of the image data *within the file*. At 4048 we start loading the image into our data array. *Note: Since, in the Random Access file mode, we must provide the address from which data is to be loaded, we define KL1 as STRT + LINX*I/2. When we start reading the data (in the nested loop at line 4050) we obtain the address of each stored data word by adding KL1 to the loop counter J.* We use the GET #3 statement to read the data from the disk (KL providing the address of the word we want to *get*, and PIXREC providing the place to *put* it. We then transfer the integer data to IMAG% and, in line 4056 we split out the two bytes and place them sequentially into our data array C(I,J2) and C(I, J2+1). At line 4068 we display this data (after setting a

Image Enhancement 273

scale factor for MAG2).

Okay, to *save* a file, we essentially follow the same procedure, except we convert the *real number* data of the array to integer words packed with two 8 bit bytes of image. First we scan through the data to find the maximum and minimum data words (lines 4106-18). We calculate a scale factor and scan through the data once again scaling the data to a 0 - 255 gray scale (lines 4122-30). We generate a name for the new file by simply adding a "D" in front of the old file name (you might want to use something more imaginative).

Now, there is a lot of "junk" in the file header that we haven't talked about, and for *our immediate* purposes, we really don't care about it; but, we must still include this junk in the header if we hope to read this file by conventional graphics software. We resort to a "quick and dirty" solution to this requirement by simply opening the original file and transferring the header to the new file (lines 4142-50).... I sincerely hope that none of you ever degrade yourselves by employing such a shabby technique. We then pack the image data (two 8-bit bytes into each 16-bit word), and save these to the disk (lines 4300-4322).

The majority of routines for this program deal with modifying the frequency domain functions (a menu is located at lines 3000-3028). Lines 3100-3120 provide a routine to filter the frequency domain function using a Gaussian response lowpass filter. The argument is found as the RSS of the harmonic coordinates (line 3106) and the attenuation is the same as the Gaussian filter used in the Gibbs phenomenon illustration of Chapter 6.

A routine to multiply the points of two spectrums (i.e., convolution) is located at lines 3200-3230. Since the data is used twice in complex multiplication it's necessary to pull it out of the array and into temporary variables (lines 3206-08) before forming the products (line 3210). A second spectrum must already be stored in arrays DAC(I,J) and DAS(I,J) and we get it there via the *Save Spectrum* subroutine which is located at lines 3400-3414 (under control of the *Modify Spectrum* menu).

Division of two spectrums (deconvolution) is performed in lines 3300-3390. This routine is written as a straightforward complex division routine, except that you may divide either the stored spectrum by the current frequency domain function or vice-versa. You must specify the direction in lines 3302-08 and a flag is set (CSFLG—line 3210) according to the selection. At line 3314 we start the actual division. Again, we pull the data from the arrays (line 3318) and then clear the array data. In line 3320 we find the magnitude, and in lines 3322-3328 we find the angle. In lines 3330-3338 we do the same thing for the stored data and in lines

3340 or 3342 we perform the division. We then convert back to rectangular coordinates (lines 3344 and 3346), putting the data back into the frequency domain array.

The routine at lines 3500-3526 truncates (i.e., zeros out) the spectrum above some selected harmonic number, using the sub-routine at lines 3900-3912 to modify the spectrum symmetrically about the Q/2, Q/2 point of the array. This is the routine we use to "cut hair" when digital noise comes up.

Finally, there's a routine located at lines 2420-2456 which will change the number of bits in the gray scale. First we find the maximum and minimum data words (lines 2422-2434), and then, based on the number of binary bits desired, re-scale and form the data into integer words (lines 2440-2452). This routine applies to the image domain, of course, and is used to experiment with the number of bits desirable for various enhancement procedures.

We have tried to show how easy these routines are to create, so we have left out the techniques used in more polished software (e.g., taking care of incorrect entries, backing out of routines when you change your mind, etc.). You can add this sort of embellishment to these programs but, in my opinion, BASIC is not a good language to write *real* software. It is, however, an excellent teaching tool, and a very good tool for experimentation.

APPENDIX 16.2

IMPROVING THE SIGNAL/NOISE & DATA RESOLUTION

This is how the story goes: A friend of mine worked on an *Extra-terrestrial Intelligence* project. Basically they have a *very* big antenna and a low noise receiver, and they search the sky for anything that looks like an intelligent signal. So much data is gathered it simply *must* be handled by a computer. The computer examines the incoming signal (i.e., noise) for anything that is statistically different from *pure* noise. Now, not wanting to miss anything just because their threshold is a couple of percent too high, they log an awful lot of noise in their archives.

So, if the computer detects something statistically *different* it starts processing the signal. It runs various algorithms and if it detects anything coherent it immediately pipes it to the control console; however, more than 99% of the stuff is quietly logged into the archives.

But the computer doesn't just tape record a lot of noise—it constantly runs tests on the data. If, for example, it detects anything remotely suggesting periodicity, it starts logging the data in *frames* (i.e., the data is logged in serial strings whose length matches the detected period). As I said, less than 1% of this stuff gets looked at; the rest is compressed and stored for "future study" (fat chance).

Anyway, I mentioned to my friend that I needed a good illustration for extracting a signal from noise and, shortly thereafter, I received a compressed file over the net with nothing more than a note saying it was logged in the direction of Andromeda. The program below de-compresses the file and displays it on the computer screen. I must warn you that, in the rush to get the manuscript ready for the printer, I never had time to actually examine this data... it's probably just garbage... but who knows, maybe you'll find the first extra-terrestrial signal....

```
' ************************************************************
' *   PROGRAM TO ILLUSTRATE THE EFFECT OF NOISE AVERAGING    *
' *      IT RETRIEVES A SIGNAL BURIED (20DB DOWN) IN NOISE   *
' ************************************************************
SCREEN 9, 1: COLOR 15, 1: CLS : PI = 3.141592653589793#
PI2 = 2 * PI: K1 = PI2 / 200: X0 = 100: YM = 170
DIM D1(4, 19), D2(2, 400), XD%(400)
FOR I = 1 TO 18
FOR J = 1 TO 4
READ A
D1(J, I) = A
NEXT J
NEXT I
INPUT "INPUT A/D DYNAMIC RANGE AS 2^N.  N ="; N
```

```
DY = 100 / (2 ^ N)
FOR M = 1 TO 60000
I2 = 0
FOR J = 1 TO 18
FOR I = I2 TO D1(1, J)
XD%(I) = INT(I - D1(2, J))
D2(1, I) = D2(1, I) + DY * INT((D1(3, J) + 141 * (RND(1) - .5)) / DY)
D2(2, I) = D2(2, I) + DY * INT((D1(4, J) + 141 * (RND(1) - .5)) / DY)
NEXT I
I2 = D1(1, J) + 1
NEXT J
CLS
FOR I = 1 TO 400
LINE (X0 + XD%(I), YM + D2(1, I) / M)-(X0 + XD%(I), YM + D2(2, I) / M)
NEXT I
LOCATE 1, 50: PRINT M
IF M > 1000 THEN 1200
FOR L = M TO 1000
FOR KTIME = 1 TO 100: NEXT KTIME
NEXT L
1200 NEXT M
1333 A$ = INKEY$: IF A$ <> "G" THEN 1333
STOP
DATA 100,0,0,0,103,0,-6,6,114,0,-1,1,118,0,-6,6,148,0,0,0
DATA 152,0,-6,6,162,0,-6,-5,172,10,-1,1,182,20,4,6,213,30
DATA 0,0,216,30,-6,6,227,30,4,6,267,40,0,0, 271,40,-6,6,277
DATA 40,-6,-5,283, 46,-1,1,287,46,-6,1,400,46,0,0
```

What we're really doing here is adding a small amplitude signal to a large amplitude noise (i.e., random numbers). As we add in more and more frames, the signal increases linearly while the noise increases in a random walk. Slowly the signal grows larger than the noise. It's interesting to note that this works even when we only have a *two-bit* A/D converter! The noise causes the bits to trip more frequently with a high amplitude signal and less frequently with a low amplitude signal. The time average then gives the correct amplitude signal when many frames of data are averaged. Try this—you can actually see things improving.

The point is: this works! So long as the noise is above the least significant bits, you can *dig out* a signal that's literally buried. Amateurs tend to throw away the noisy bits—thereby excommunicating themselves from the wondrous things that lie just outside the immediate view.

BIBLIOGRAPHY

To my way of thinking, it makes no sense at all to fill up pages with long lists of references that are obviously not appropriate for the intended audience. As noted in my previous book on the FFT, the following references require a knowledge of the calculus. If you want to pursue this field, but are unsure about your ability to handle calculus, I recommend Silvanus P. Thompson's *Calculus Made Easy*. Widely available in book stores, this 1910 publication can be read in a week (two at most). If you can read and understand this little book you can probably survive even the worst college presentation. With these comments in mind, I recommend the following:

Bracewell, R.N., *The Fourier Transform and Its Applications*, McGraw-Hill. This is my standard basic recommendation. Dr. Bracewell is well known for his various accomplishments, but we know him best for pushing this stuff a little farther across the table (so those of us with shorter arms might have a better chance of reaching it).

Brigham, E.O., *The Fast Fourier Transform And Its Applications*, Prentice Hall. This is another good reference for the FFT. Its strongly graphical presentation (especially of the 2-D FFT) is helpful.

Gaskill, J.D., *Linear Systems, Fourier Transforms, and Optics*, John Wiley & Sons. If you are interested in pursuing Fourier Optics this is pretty much the standard text. The tutorial orientation is especially helpful and the *review* of fundamentals is valuable if you have been away from this material for some time.

Hubbard, B.B., *The World According to Wavelets*, A K Peters, Wellesley, MA. This is a very readable introduction to wavelets; although, much of it is oriented toward the mathematically savvy.

Walker, J.S., *Fast Fourier Transforms*, CRC Press. This book has a lot of good material, but (for those who can handle the math) I especially recommend the introduction to Fourier Optics (Chapter 6).

INDEX

A/D conversion 111, 140
Addition theorem 128, 129
Aliasing 65, 110
Area 26, 27
Argand 4, 9, 10
Argument 3, 5, 8, 221-224
Array 70, 71, 73, 93-97, 158, 159, 209, 219, 226, 233
Average 25-29, 75, 137, 138, 159, 160, 166, 167, 185, 199

Band-limited 47, 117
BASIC 64, 104, 155, 162, 163, 205
Basis function 36, 104, 182
Bessel function 171, 173, 260-262
Binary 137-139, 193
Bit reversal 52, 77, 162
Butterfly 162

Calculus 7, 200, 201, 277
Camera 158, 187, 191, 193
Capacitor 14, 35, 41, 59
Circuit analysis 31, 34, 37, 64, 66
Coefficient 8, 18, 43-45, 87, 128, 129, 160
Complex addition 19, 22
Complex conjugate 10, 22, 23, 25, 26, 68, 70, 221-223
Complex exponential 21, 24-29
Complex multiplication 21-24, 64, 273
Complex numbers 4, 10, 259
Convolution 21, 24, 31, 58, 63, 101, 111, 133-135, 185-196, 205, 273

Deconvolution 185-196, 273
DFT 1, 2, 26, 50, 68, 72-77, 83, 92, 104, 113, 117-118,
                   126, 128-131, 157, 158, 201, 203, 206, 209, 252
Differentiation 15, 16, 196, 201-204
Digitization 110, 111, 138, 140, 142
Discrete Fourier transform 1

Eigenfunction 15-17
Execution time 64, 163
Exponential 6, 7, 10, 17, 219

Fast Fourier transform 70, 103, 211, 277
FFT 1, 2, 50-56, 66-82, 91-97, 104, 116, 158-165, 179-184,
                 209-210, 211-212, 215, 220, 223, 225, 227, 229, 277
Filter 46, 58-64, 119, 123, 185

Frequency domain 24, 31, 35, 58, 60, 65, 66, 97, 114-116, 124, 126-128,
    129, 131, 145-147, 160, 163, 170, 173, 181, 185-189, 192, 196, 201-204
Frequency response 58, 63, 64, 66, 121-123
Frequency shift 145-147, 152
Functions 1, 12, 16, 22-23, 28-30, 36, 91, 97-98, 126-131
    159, 170-173, 182, 201, 258-262
Fundamental 16, 29, 93, 95, 125, 127, 128, 156, 159, 202

Gaussian 60, 62-63, 103, 188-194, 271, 273
Gibbs Phenomenon 48-51, 56-57, 64, 125-126, 170, 173

Harmonic 47-50, 56-57, 61, 63, 95-97, 100-102, 112-116, 118, 122
    142, 145-147, 154-157, 166-168, 179-181, 196, 201, 203, 223

Imaginary number 4, 36
Impulse function 66, 112, 114, 126, 128-130, 172-173, 179, 181, 186
Inductor 34-36, 41, 60
Integration 16, 26, 196, 201-203, 204
Inverse DFT 50, 83
Inverse FFT 83-90, 154

Laplace 17
Limit 28, 34, 47, 49, 109, 119, 123, 173, 187, 193
Linear system 16, 186, 201
Logarithms 214, 220

Maclaurin series 8
Magnitude 6, 7, 21, 23, 26, 98, 160, 168, 179, 180, 213, 257, 273
Modulation 19-23, 69, 99-100, 111, 116, 147, 149, 157, 181, 182

Negative frequency 56, 68-70, 82, 83, 118, 119, 124, 146, 147, 160,
    164, 167, 169, 205, 221-224
Negative time 222, 223
Nyquist criterion 110
Nyquist frequency 61, 74, 77, 106, 107, 118-120, 164, 221, 224
Nyquist sampling rate 105, 106, 224

Orthogonal functions 25-28, 30, 93, 101, 102, 104
Oversampling 123-125, 131-133, 145, 173, 188, 225-227
Phase shift 19, 36, 58, 128, 147, 181
Phasor 4, 9, 18, 19, 21, 23, 26, 83, 86-88, 124, 160, 223
Polynomial 60, 201

Real function 128
Real part 45, 46, 87

Reconstruction 93, 96, 114, 132, 133, 149, 152, 160, 165, 173
Resolution 38, 47, 56, 97, 102, 104, 142, 184, 187, 192-195

Sampled data .. 2, 97, 100, 107, 110, 111, 117, 118, 120, 133, 134, 146, 224
Sampling 69-70, 93, 102, 105-120, 140, 142, 148,
164-165, 221, 223, 224, 233
Scaling 16, 28, 49, 115, 154
Shifting theorem 126, 145-147, 152, 212
Similarity theorem 102, 145, 173
Sinc function 96-103, 130-134, 146, 149, 169, 170, 173, 184
Spectrum Analyzer 91-93, 95, 96, 101, 132, 140, 142, 212
Square wave 17, 47-50, 63-66, 99, 112, 139, 202-205, 223
Stage 67, 72, 75, 77-80, 83, 84, 88-91, 91, 162, 212, 229
Step function 51, 56, 63, 64, 205
Stretching theorem 68, 70, 114, 152, 208, 211

Taylor series 67
Time domain 56-58, 83, 103, 115, 124-131,
151-154, 170, 181, 185, 187, 196, 202, 223, 226
Time shifting 145, 147
Transfer function 37, 46, 58, 60, 64, 66, 186-188, 190
Transient analysis 31, 64
Truncation 137, 138, 191, 195
Two-dimensional FF 158-161, 168, 170

Vector 9, 21, 26
Waveform 19, 20, 25, 100, 114, 115, 153, 154, 223, 226
Weighting function 103, 132, 143, 149, 219, 220, 227, 228, 233